bush
PUBLISHING
& associates

Pull Each Other Up

How women in STEM can work together to pull each
other up to higher heights and inspire future generations

HEATHER GRAHAM

bush
PUBLISHING
& associates

Pull Each Other Up: How women in STEM can work together to pull each other up to higher heights and inspire future generations

ISBN (paperback) 979-8-9854481-2-2
ISBN (ebook) 979-8-9854481-3-9

Printed in the United States of America.

First printing 2022 by Bush Publishing and Associates, LLC
Tulsa, Oklahoma
www.bushpublishing.com

Bush Publishing & Associates, LLC books may be ordered at bookstores everywhere and at Amazon.com.

Cover Art, Layout and Design by Bush Publishing and Associates, LLC, Tulsa, OK.

Editing by Writing by Michele, LLC.

All opinions expressed in this book are the author's.

DEDICATION

For my girls. And my girl's girls.

TABLE OF CONTENTS

ACKNOWLEDGMENTS

The stories in this book were gathered from real women in the working world who face biases and stereotypes every day. I would like to thank all of these women who were willing to share their thoughts and stories with me through personal interviews and survey responses. Here's to making it better for all of us!

Chapter 1

STEM IS OUR FUTURE

"Can you hear me?"

How often have we heard that phrase uttered on Zoom and conference calls, before and even more during the Covid pandemic? People forget that they are muted, until they finally find that right button to join the conversation. Or, they try to jump in with something to say but get cut off. Their words may be a way to interrupt the conversation, or they may be simply seeking validation that people will hear their important statement.

This is how women in STEM feel every day. Incredibly smart, talented women feel they are somehow muted because why else would no one hear what they are trying to add? Or even worse, they are just being ignored, whether deliberately or accidentally. Why is this still happening? We have diversity programs, and we tell girls that they can do anything, yet they are, overall, not choosing STEM.

What is keeping them away?

Women could rule the world…if we wanted. If the world was suddenly thrown into a catastrophic situation, and the only way it could be saved was for women to take over, you better believe we would step up. Everyone would have safe shelter and food. The kids would even still have school. As Margaret Thatcher once said, "If you want anything said, ask a man. If you want anything done, ask a woman."[1] If you have ever seen women organize a fundraiser for a family down on their luck, or coordinate a post-funeral lunch or even a kid's birthday party, you know we are fully capable of organizing and leading. Women tend to be empathetic, team-oriented, and find solutions that are good for the group rather than the individual. With all of the diversity and inclusion programs in place, you would think these traits would be desired and even specifically looked for in recruiting. Instead, they are often seen as weaknesses, as reasons to pigeonhole women into lesser roles.

Over the years, women have faced overt discrimination in many male-dominated fields like Science, Technology, Engineering, and Math (STEM). Men have not been afraid to tell women that we have no place in their world. A coworker told me this story from her college days:

My sophomore year in university, after I'd taken all of the prerequisite classes and earned As and Bs, I was accepted into the Electrical and Computer Engineering program. My first class in the program was taught by a man who started the first class by saying, "Women can't be engineers. All of the women in this class need to drop and change majors." I stuck with the class, but he gave me a failing grade. I retook the class with another professor, passed, and went on to the next class in that major, where he was again the professor. Again, he started the class by saying, "Women

---

1 "The Tough Top Tory in Britain Wants to be 'Madame P.M.'" *People*, 15 Sept. 1975, p. 11.

can't be engineers. You need to drop," and failed me again. This particular professor was tenured and last I heard, he has an entire filing cabinet dedicated to his EOO complaints. But he's still teaching—and I ended up changing my major.

Even though she had all the evidence to prove that she could understand and do the work, this teacher was bound and determined to keep her out of the field simply because she was female. And unfortunately, it worked.

I joined the traditionally male-dominated field of technology right before the Y2K boom, when companies were screaming for people to change all of the old computer code so it could handle the four-digit year, thus avoiding catastrophe when we flipped from 1999 to 2000. I was one of only a couple of women in any of my computer science classes, one of the few of us graduating with a computer science degree. Since then, most of the people I worked with and for have been male. I remember my college advisor being so excited that I was going for a computer science degree, specifically because there were so few women in the program, and that was only twenty years ago!

Have things changed? The numbers show change, but women in STEM still face many obstacles that are keeping them away from entering, or from staying and rising in, their field. More women than men now earn degrees in the STEM fields through the Master's level, which is good news, but these numbers drop at the Doctorate level. Still, the number of women earning these degrees has leveled off or dropped.[2] Even though more women are earning technology degrees, in the United States women make up only 25 percent of technology workers, while 80 percent of technology executives are men. Unfortunately, that's not much of a change from what I saw in 1999.[3]

What is happening here?

The problem is, while more women are entering the field, almost 56 percent leave by mid-career, which is double the number of males who leave.[4] Fewer women in the field means less to promote, and fewer mentors for the next generation. Even though women are changing the stereotypes around interest in STEM fields, the view remains that there are not as many women in the field and in leadership, making it difficult to change those stereotypes.

We know there is also a huge gap in women leading organizations. In the United States, women make up 47 percent of the workforce. We hold about 52 percent of the management and professional occupations. We show up strong in fields such as medicine, human resources management, and social services management...at least until we get to the leadership level. Sarah, who works in healthcare administration, told me: "My organization has over 3,000 employees with only 376 men. Most of

2 Sue V. Rosser & Mark Zachory Taylor. "Why Are We Still Worried About Women in Science?" *Academe* 95, no. 3, (May-June 2009): 7-10.

3 Anjali Shaikh, Kristi Lamar, Kavitha Prabhakar, and Caroline Brown. "Repairing the Pipeline: Perspectives on Diversity and Inclusion in IT." *Deloitte Insights*, 15 Oct. 2018. Web. https://www2.deloitte.com/us/en/insights/focus/cio-insider-business-insights/perspectives-on-gender-diversity-and-inclusion.html

4 Deborah J. Armstrong, Jason E. Nelms, Cynthia K. Riemenschneider, & Margaret F. Reid. "Revisiting the Barriers Facing Women in Information Systems." *Journal of Computer Information Systems*, 53, no. 2, (Winter 2012): 65–74.

the men are in leadership positions." So 10 percent of the staff are men, yet they are the majority of the leaders. How did it get so skewed?

And this trend occurs everywhere. The number of women in leadership positions across all fields for women is dismal. For example, the highest level of management is usually executive, specifically those "Chief Executive of..." titles. When it comes to the higher levels of management and leadership, women only make up 28 percent of those Chief Executive spots.[5] In 2020, in the United States, we had the highest number of Fortune 500 companies led by women in history—*thirty-seven*. That's thirty-seven companies out of five hundred, which is less than 10 percent. And you know those companies employed more than 10 percent women on their staff. This is still a much better statistic than twenty years ago, when there were only *two* Fortune 500 companies with women CEOs.[6] Where did we go wrong in showing our leadership talents as a benefit to the professional world?

It starts at the bottom. The jobs of our future in the United States are in the STEM fields, and there is already a huge shortage of qualified workers in these industries. Cutting out an entire gender does not just hurt women, it hurts our companies and our future growth. For example, in 2020 an estimated 1.4 million technical jobs were open in the United States but only about 29 percent could find qualified graduates to fill them.[7]

---

5 "Labor Force Statistics from the Current Population Survey." 22 Jan. 2021. U.S. Bureau of Labor Statistics. Web. https://www.bls.gov/cps/cpsaat11.htm.

6 Hinchliffe, E. (2020, May). "The number of female CEOs in the Fortune 500 hits an all-time record." *Fortune.* Web. https://fortune.com/2020/05/18/women-ceos-fortune-500-2020.

7 Kristin Holmberg-Wright & David J. Wright, "Why Gender Diversity is Both a

Along with fewer women staying in STEM fields, we are also being advanced at much lower rates than men, with only 18 to 20 percent of executive positions in technology organizations being filled by women.[8] When you have fewer women joining, there are fewer to promote, creating that familiar, built-in excuse of lack of female leadership talent available for those roles. As recently as 2018, I looked at the technical organization chart of a company I worked for. Because it had a female CIO and CEO, I thought perhaps the numbers would be better, but only 11 percent of the management positions were filled by females. This was a company that I greatly admired for putting so much attention upon diversity and inclusion, yet still, the numbers were that skewed! On top of it, women of color fare even worse in the professional and managerial fields, with less than 15 percent in management positions in all fields, and only 4 to 6 percent in the Chief Executive role.

Why are women not joining or staying in these lucrative fields, and why are they not being promoted to these leadership levels? What are they seeing in the field that causes them to lose interest? Much of the overt sexual discrimination is on the decline in the United States, as the public, in general, does not accept this behavior and people are more empowered to speak out against it, but the behavior remains engrained in our culture and workplace. In today's world, most people and companies realize they cannot allow any blatant sexist or discriminatory behavior, or they open themselves up to lawsuits, and perhaps

Challenge and an Impending Financial Growth Opportunity for the Global Technology Industry," *Business Education Innovation Journal*, 10, no. 1, (June, 2018): 51-58.

8 Jodi Helmer. "Breaking the Tech Ceiling." *University Business*, 20, no. 8, (August 2017): 43–45.

even worse, reputational risk. With all of the social media sites and open sharing of opinions, an organization that has a nasty, cutthroat atmosphere will find itself with negative reviews on sites such as Glassdoor and LinkedIn, where many professionals go to research new opportunities and organizations. As Shauna, who has worked in healthcare for several years, notes: "Due to change in times and also to laws (inclusion and diversity), there is more pressure to treat people equally, and people generally follow the rules. But behind the scenes, the judgment is still there. Stereotypes and mistreatment still exist. People can be very fake." The feelings are still there, under the surface.

This is not to say that there are companies or areas where overtly discriminatory behavior is tolerated as part of the deal if you want to get ahead. But that attitude is some of the exact reason fields such as STEM are struggling to attract and retain a diverse workforce.

Many companies recognize there are problems. There have been great strides in diversity programs and a push to bring in more talent regardless of race, gender, sexual orientation, religion, or any other dimension a person may be known by. These programs started with a focus on race but expanded in the 1990s to include gender, and today there are many more diversity and inclusion dimensions considered. Organizations offer training and education to increase sensitivity to the barriers that women (and others) face.

The problem is, these programs are slow to have an impact and are not changing the face of the organizations. Some even feel that highlighting a group in diversity training means that the group is lesser and needs special treatment to succeed, which only furthers stereotypes—not the original goal of the training! A 2018 study done by

Deloitte with CIOs in the technology industry found that 44 percent of these technology organizations had no specific diversity and inclusion hiring practices, education or training, or retention practices. Worse, even among those that have existing diversity programs in place, only an average of 32 percent of staff are women, which is only 10 percent more than those without diversity programs.[9] Diversity programs are a start, a good attempt to make a difference; however, there is more to the story.

We can remove all of the external problems and overt discrimination, but if we do not change our inherent thoughts about how women should be or act, then we have not made any real progress. It is not enough to chase overt sexist behavior into hiding while still letting it creep around the dark corners. We need to look deeper into our behavior toward ourselves and each other to find where the problems still exist by looking at the stereotypes, biases, judgments, and microaggressions women still face in the STEM fields and how it is impacting the numbers of women joining and staying in the field. We will get real about what women are going through by looking at facts and statistics that have been reported and what actual women are still experiencing in the field. We'll also consider the ways people think about, talk about and treat each other without even thinking about the impact of their actions, as well as the thoughts and doubts that keep women from standing up and excelling wherever or in whatever we want.

---

9 Anjali Shaikh, Kristi Lamar, Kavitha Prabhakar, and Caroline Brown. "Repairing the Pipeline: Perspectives on Diversity and Inclusion in IT." *Deloitte Insights*, 15 Oct. 2018. Web. https://www2.deloitte.com/us/en/insights/focus/cio-insider-business-insights/perspectives-on-gender-diversity-and-inclusion.html

Finally, we will also look at some solutions that can help bring some of the problems to light and transform people's actions and behaviors. We need real talk and real solutions to change what women in STEM face if we want to attract and retain the best talent. This is the only way to change the status quo and make sure we are being heard.

Chapter 2

STEREOTYPES

There are so many old and just plain wrong beliefs that hold women back. Let's start with stereotypes. Where do stereotypes come from? *The Merriam-Webster Dictionary* defines the word *stereotype* as "a standardized mental picture that is held in common by members of a group and that represents an oversimplified opinion, prejudiced attitude, or uncritical judgment."[10] Stereotypes are the way large groups of people think or view another group of people, based on what everyone else in their group believes. It is the act of bucketing people. We like to put people into buckets and sort them by color or gender or how you dress or how you speak…and the list continues.

The problem is, there are only so many buckets. So, we put people where we *think* they fit, even if that bucket is full or the wrong size for that person! Stereotypes, in other words, are a quick way for us to make a judgment about someone without having to take the time to get to know that person. They are a lazy way to sort and categorize others. Bucketing does not take into account any sort of individualism,

---

10 "stereotype." *Merriam-Webster.com*. Merriam-Webster, 2011. Web. https://www.merriam-webster.com/dictionary/stereotype

and historically, these categories have represented the lowest common denominator to catalog people and hold them down. Once a person is defined in a certain way in people's minds, that is where they stay. And it is very difficult to change a mind.

Are stereotypes untrue? Yes and no. These buckets or stereotypes may have some basis in fact for some subset of a group, but they are generally applied to everyone in the group rather than seeing people as individuals.

For example, a common stereotype for women is that we are the weaker sex and that we would not want or could not handle a job that involves physical labor. Yes, studies have proven that men have overall more skeletal muscle than women even when adjusted for mass differences, which means they tend to be physically stronger.[11] Men also tend to be larger and therefore have more muscle overall; this contributes to greater strength among men, although women such as Abbye Stockton, a powerlifter in the 1930s-40s, and one of the first recognized women powerlifters at only 5'5" and 115 pounds, would disagree.[12]

The issue with accepting at face value the fact that physical strength tends to be greater in men than women is that it ignores outliers such as Stockton and focuses only on the average. It also has been extended to mean women are weaker overall, not just in physical strength but in all things—and therefore it's assumed we cannot handle intense mental or psychological situations. Not long ago, women were thought to

---

11 Walter Watson et al. "Predictive Equations for Skeletal Muscle Mass." *American Journal of Clinical Nutrition*, 73, no. 5, (May 2001): 993-994.

12 Miller, Tom. "The 10 Strongest Women to Ever Walk the Earth." *Fitness Volt*, 19 March 2020. Web. https://fitnessvolt.com/strongest-women/

need a man to protect them physically and mentally from the challenges of the world.

Physical strength may have been a requirement historically to fight off predators and drag home the day's kill to provide for their mate and family, but we do not live in caves anymore. Even though men may be stronger physically, we now use tools and machines to help us do physical labor. The whole field of STEM is about using technology to make our work and our lives easier, and therefore less physical. Does it really matter, then, if a woman is physically not as strong as a male counterpart? Does she not bring a unique and valuable perspective into how to make things easier to do?

What about the stereotype of women's "delicate constitution"—being weak, fragile, or having a frail nervous system that is unable to handle the stress that can come in the intense STEM fields? Any woman who has gone through pregnancy and birthing or any sort of intense life challenge can testify that women are far stronger at handling stress than we are given credit for. Scientific studies have shown that women's brains do handle stress differently than men's. Women tend to handle chronic stress better, while men tend to handle acute stress better. Men also remember stressful events as an overall picture and feeling, while women will remember finer details. Contrary to the popular stereotype of the hysterical woman in a stressful situation, brain studies show that women can control their emotions better than men![13]

---

13 Cahill, L. (2009). "His Brain, Her Brain." *Scientific American Mind, 20*(3), 40. https://doi.org/10.1038/scientificamericanmind0509-40

What the studies show is that while biologically, we have differences in how we handle stress and show emotions, this does not infer that women are mentally weaker. These results simply reveal that we need to consider potential differences in how women may view and react to situations versus men. In fact, when faced with a difficult problem, it would be better to have varying methods of dealing with stress and seeing the problem in order to find innovative ways to solve it!

In addition to muscle size, men's brains also tend to be larger than women's. This fact has unfortunately led to an unfounded stereotype that women are not as smart as men. Some still believe that women are somehow intellectually inferior and therefore not intelligent enough for STEM fields. Even as recently as 2005, it was stated by the then-president of Harvard University that anatomical differences in women's brains might be the reason for fewer women excelling in the math, science, and engineering classes—an excuse for why there was lower enrollment and retention of female students in those classes.[14] Instead of looking at reasons why women might feel uncomfortable in these classes (such as being told they are not smart enough for that program!), lower intelligence is given as a reason they leave. It is ridiculous that these thoughts are still being expressed in the twenty-first century in a center of higher learning.

While differences in the size and structure of certain brain areas seem to give credence to the concept that stereotypes do start somewhere, when you compare relative sizes to the overall size of the brains

14 Ibid.

of men versus women, they are equal.[15] Even taking into account how brain sizes can vary, studies have shown there is no difference in general intelligence between men and women. There are slight variances in specific areas of ability related to intelligence. Women tend, on average, to be a little better on tasks requiring verbal ability such as knowledge of language or pulling together and understanding information from different sources, while men tend to be a little better at spatial ability or manipulating objects either physically or mentally.[16] These differences mean men and women look at problems from different angles and process information in different ways, which should be celebrated and encouraged in the business world to find new and creative ways to solve problems, not to inhibit the opportunities of a whole gender.

The fact that a former president of Harvard University was looking at brain sizes to reason out why fewer women are in the STEM fields points to another stereotype—the idea that women do not have the aptitude or desire to study STEM. If women are less smart than men, why are more women getting college degrees? It would seem that the contributions of women such as Marie Curie, physicist and chemist, and pioneer of research in radioactivity; or the contributions of the human "computers," the group of women in the 1940s-50s who did manual mathematical calculations to determine the proper trajectories for launching the first space rockets; or all the women who contributed and even created computer programming languages are seen as anomalies. Unfortunately, these women were undervalued during

---

15 Bishop, K; Wahlsten, D (1997). "Sex Differences in the Human Corpus Callosum: Myth or Reality?" (PDF). *Neuroscience & Biobehavioral Reviews*. 21 (5): 581–601. doi:10.1016/S0149-7634(96)00049-8. PMID 9353793.

16 Hunt, Earl B. (2010). *Human Intelligence*. Cambridge University Press. p. 389.

their times and were not in the spotlight getting credit for their work until much later.

Instead of asking why there are not more examples such as these women who had no trouble with intelligence or ability in these fields, we have people trying to find physical reasons to show that women just are not suited for STEM. Why not consider that a comment such as that from a public figure at a large, respected institution might be the reason girls and women believe the stereotype and do not even try to join the field? Or that these comments contribute to the stereotype continuing to be believed by both men and women?

This goes to the root of the problem with stereotypes; they are very pervasive throughout our culture and thoughts. People will believe them to be true and see facts to support their beliefs because "everyone knows it's true." They will take an example that fits the mold of their belief and point to it as proof, or take an untrue statement as fact to support the stereotype as true.

Chapter 3

GENDER STEREOTYPES START YOUNG

Where do these stereotypes start? As early as grade school, girls are subjected to this stereotype that they lack aptitude in math and science. Girls and boys in elementary schools show similar test scores on standardized math tests, yet the girls feel less confident in their math abilities because some teachers still describe the male students as better in math than female students. This view expressed by a trusted adult sticks in students' brains, both females and males, and sets the stage for fewer females to explore higher levels of math and science in high school and college. By then, they do not pursue certain career fields such as those in STEM because they feel they do not have the innate ability to do the level of math needed, regardless of what the actual test scores show.[17] Children even describe science and math careers as more masculine fields.[18]

---

17 Anke Heyder, Ricarda Steinmayr, & Ursula Kessels. (2019). "Do Teachers' Beliefs About Math Aptitude and Brilliance Explain Gender Differences in Children's Math Ability Self-Concept?" *Frontiers in Education*, 4. https://doi.org/10.3389/feduc.2019.00034

18 Dale Rose Baker. *Understanding Girls : Quantitative and Qualitative Research.* Boston: Sense Publishers, 2016, ProQuest eBook Central.

Girls also face teasing when they show their intelligence. We have been told over the centuries that women should be demure. To be too smart means that you will end up alone and not find a spouse. Women have been expected to run the house, be a good wife and serve her husband. Intelligence is a challenge to the status quo. Unfortunately, some of these feelings persist and are subconsciously getting fed down to our children. Times are changing, as more children think of women in roles such as scientist or astronaut, but more needs to be done. Girls should not feel they need to hide their intelligence; rather, it needs to be celebrated in everyone, in all our various skills and talents. We need to change the stereotype that a smart girl interested in STEM is a "geek" or "nerd" who is a loner with no friends outside of their computer or lab. Girls should not be afraid to show their intelligence or interest in STEM so they can fit in. We must make it cool to be smart! As Emma Watson is quoted as saying, "girls should never be afraid to be smart."

Another area where stereotypes arise is in demeanor. Girls are often expected to be quieter and more submissive than boys. If a little girl shows competitiveness or assertiveness, she is called "bossy," which is discouraged, while the same attributes in a boy are seen as positives, and thus are encouraged. On the other hand, if a boy is not as assertive or outgoing, they may be called "pussy" or "weak" or "girly," slurs insinuating that these stereotypical female traits are not as good as male traits. Boys may hear that they should not cry like a girl, while girls hear they are hysterical if they show too much emotion. These messages emphasize that going against gender norms is considered wrong, which sets the stage for similar interactions later in life. It also contributes to an overall lack of confidence among girls, and a reluctance to speak up or express opinions and emotions. This can affect potential leadership opportunities later too, as showing confidence and speaking out are part of being seen as a leader.

Another significant stereotype that starts early and impacts many women is the idea of beauty being important to a woman's career. Studies have shown that "physical appearance is more important for females than for males because the culture values an attractive appearance more in females than males."[19] From an early age, boys and girls (aged 3–10) are already aware of gender stereotypes, especially those related to female appearance in the United States, such as the idea that being pretty means wearing dresses and using makeup.[20] If a girl does not want to wear dresses or the color pink, she is called a "tomboy," which reaffirms the thought that she is not behaving as a girl should.

What do these types of stereotypes mean for women? Gender stereotypes not only describe how we feel a person should look, feel and act, but they also convince us that is how someone of a particular gender should act—and if we do not fall into these norms, we are somehow wrong or broken. The more people, male or female, hear these stereotypes, the more they believe them and expect others to act accordingly. If a girl is taught that she should be submissive and not speak up, she will more than likely follow those rules throughout her life and not be perceived as a leader. If on the other hand she is not submissive or quiet, others will comment that she is not being "ladylike."

19 Jackson, Linda A. 1992. *Physical Appearance and Gender : Sociobiological and Sociocultural Perspectives*. SUNY Series, the Psychology of Women. Albany: SUNY Press

20 Neil Howlett & Karen J. Pine & Natassia Cahill & İsmail Orakçıoğlu & Ben (C) Fletcher (2015). "Unbuttoned: The Interaction Between Provocativeness of Female Work Attire and Occupational Status." *Sex Roles* (2015) 72:105–116

If a child is told and thus believes that males are better in math and science, then they will feel that girls will struggle and not do as well, regardless of what the scores show. They will neglect to offer encouragement or help because "she won't understand it anyway." If the girl feels like a failure because she has been told that she will not be as good at the subject due to her gender, she may decide to go a different route with her studies instead of trying to get more help or seeking to understand what she is interested in. Looking around at different job types, if people do not see others like them in roles, then they may feel like an outsider and not have an interest in those areas.

This concern with behaving outside the stereotypes or norms seems to peak around middle school for girls. Many adults will remember and understand how important it was (and is) to fit in during the middle school years. Girls start indirect aggression early, around that adolescent/middle school age. They are taught that aggressive behavior is not preferred in females, so as part of the attempt to establish their level in society, they find other ways to put other girls down. And girls are more aggressive toward other members of the same gender than males, even at this age. They engage in verbal attacks, criticism, and indirect aggression by gossiping instead of physical fights.

Girls and boys are bombarded with media, popular culture, friends, and families telling them how they should look, act, and behave to be part of the popular crowd. Unfortunately, this means that girls decide topics like math and science are no longer "cool" and will tease other girls that continue to show interest in them. And this is where we lose many who had interests in STEM. They might have been very drawn to these topics in grade school, but the pressure from others during this time of growth and change drives many girls to turn away from those interests so they can instead fit in. They do not want to seem uncool to their friends, and it is around that middle school time when the "nerd" term arises as a negative. As a result, many girls shy away from these areas, even though we are such a technology-based society and the use of technology has never been higher. Middle school is therefore a very important stage to not only engage and encourage girls to stay interested in various STEM topics, but even more importantly present them in a way that makes them a cool thing for anyone to do.

There are great programs to involve girls in traditionally male-oriented fields such as STEM, and they are showing up at younger ages in schools and clubs. The challenge is how to present the programs in ways that interest girls. This means highlighting women who work in these fields, having women collaborate with these programs and the girls and relay their real-life experiences, and designing the topics in a way that appeals to girls, especially where their interests lie at each age. Girls do not want to be seen as antisocial, awkward geeks when trying to fit into their peer groups, yet unfortunately, the perception of most in STEM fields is exactly that. To counter this stereotype, we need to show many different people who work in these fields so girls can see themselves in someone. Until we can change the perception of a field or topic to become inclusive and inviting to everyone, and show women working on and enjoying the career right along with men, it will be hard to convince girls who are in this stage of finding themselves and situating themselves with their peer groups to go against the grain and do something seen as "not cool" for girls.

We need to find ways to meet girls where they are, get them involved, show them why STEM topics are interesting and relevant to them, and continue to support them through high school and college to take classes to develop those skills. Show them that they have the intelligence for any field and that, while it takes effort, they can do what is needed to succeed in STEM fields. Bring female role models in, and change the language and image that people have around these areas. Create social groups of girls interested in these topics so that no one feels alone. We need to get in early and change minds about fields considered male-only so that we help fill skills shortages with girls who formerly would not have considered going into them. If we lose their interest at this stage, most will not come back later and will miss opportunities for the education needed to enter these fields.

We need good female role models in the STEM areas, showing off their intelligence and abilities so girls and boys both can see that women in STEM are not loners and geeks, but valuable members of the field. And we need to keep these role models relatable, a normal part of the team. The media tends to sensationalize or exaggerate women in these roles. If we always show women in STEM as the lone wolf who is either overly eccentric or the beauty queen turned scientist, it still does not show that women are simply part of the team, just like the men. We like strong role models who show they are smart and tough, but we do not want to be seen as the outlier or the one team member who checks the gender box. We can highlight the teamwork and communication abilities, the relationship building, and the normal woman who happens to work in STEM. The more girls can see themselves in others in the profession, and the more normal we make seeing women in the profession, the better the chance that girls will want to continue in their STEM-related interests.

Chapter 4

WOMEN ENTERING CAREERS

What happens when these girls start to enter the workforce with these preconceived notions of how they need to be and act? Women are often pushed into roles that are stereotypically more "female," such as nursing, teaching, and social sciences where the warm, nurturing personality expected of women is seen as a good fit. Jackie, a teacher and school principal from England says,

People expect women to be teachers, as it is a stereotypical profession for women. When I was in high school, our guidance counselor (old school and slightly sexist) advised me that I would be best suited to teaching. When I asked why, it seemed to be that this advice was largely based on my gender and not on my grades (within the top 10% of the class) and certainly not my interests.

When women do enter traditionally male-oriented occupational fields, they are often expected to take the supporting roles. Women are expected to be the nurse, not the doctor. Or the paralegal, not the partner of the law firm. When a client or customer contacts the organization and reaches a female, they ask for the male in charge.

With the encouragement to go into traditionally feminine fields, many women become less confident in their abilities to perform more typically male-dominated careers such as STEM—which leads to a vicious circle of fewer women in those fields.[21] This is not due to lack of desire, but when females are told they are better suited for other areas of work, they grow up less confident about their ability to join and excel in traditionally masculine careers, and there are fewer women in those fields to prove the stereotype wrong. Even women do not expect to see other women in fields such as technology. As a cybersecurity engineer puts it, "I'll admit that I often assume a woman isn't technical and will adjust the level of detail based on that assumption. It's always nice to be pleasantly surprised when the woman shows she knows her stuff."

Women entering the STEM field may have a hard time fitting into the workplace and getting advanced. Many companies still have that "bro culture" or "good old boys network" that keeps men involved in internal networking, exciting and high visibility projects, and social events, while excluding women from hiring, promotions, and decision-making circles. We still see some more overt discrimination in certain areas; for example, women are often expected to get the coffee and take meeting notes in professional settings, while also keeping quiet. In such settings, people do not think the woman is there to lead the meeting, and will relegate her to a side role, assuming she is there as an assistant or to take lunch orders. Often, people may refuse to speak to a woman if a male is available instead.

---

21 Coffman Katherine Baldiga, "Evidence on Self-Stereotyping and the Contribution of Ideas." *The Quarterly Journal of Economics*, 129, no: 4, (2014) 1625-1660.

In addition, some people of both genders just do not like seeing a woman in power or being powerful in any role! Lora, who works in state government, says she cannot advance because she has not worked in government for twenty-plus years, and it's not what you know but who you know to get information about promotions and opportunities. Furthermore, she says, "I want to quit and do something else, but at my age, I just stay—at least for now..."

Such sentiments are not pouting because we are not being included or our role and title acknowledged. We know that to get noticed, networking and face-time with those in charge are key. This means women are missing out on important opportunities by not being included in these events. Brenda, a woman with thirty years in sales relates, "I do know as far as 'bonding', I missed out on that part with the male management simply because I was a woman. Bosses would engage in after-hour activities with male salespeople." This behavior is not only off-putting to women, it hurts the desire to stay and advance in that company or department. Furthermore, Brenda says "I've worked at two major companies in the past twenty years. Both were strong believers in the 'Boys Club'. Advancements into management almost always went to males even though there were women more qualified. I think it goes back to fraternizing outside of work. Golfing, drinking, etc...."

In addition to trying to fit in with the boys club, women face criticism from both sides if they do not fit the part as expected down to dress, hair, and makeup. The gender stereotype about attractiveness in women carries into adulthood. Women at all levels are expected to be more attractive and dress nicer than men, even in workplaces or departments that are not public-facing. According to Tina, an IT professional for over twenty years,

We are expected to dress nicer, act more professional... Tattoos are looked down on if they are on women but not on men. Appearance, in general, is expected to be better on all levels. I was frowned upon by my senior supervisor for having a tattoo that was covered 99 percent of the time. However, he had no problem with a man who had an entire visible sleeve of tattoos.

And this is in a field that usually has a more casual dress code in general and looks have no relevance to the job.

As a woman matures in her field in STEM, she will often see more men advanced than women. When questioned about any disparity, organizations will try to rationalize their hiring and promotions. Brenda relates a situation that she experienced:

I was working at a company and moved from a sales position into a position I created myself. The next step was regional manager. There was a man that they had brought in from outside the company to fill one spot. They listed the salary for the position, and I applied. When I interviewed, the hiring manager stated that my pay would be $40K instead of the $50K that the man was just given. I flat out said that isn't acceptable and it's discrimination. They tried to say that he had more experience than I did. I quickly put that one to rest. If they only looked at my time with their company, it would appear so. However, I pulled out my resume and stated that I clearly had more experience than him. They then told me that they had eliminated my previous position and if I didn't take this one at the rate they offered, I would have to leave the company. I took the job.

I had a similar situation in that a male was hired to do the same job as myself, and the hiring manager had been his friend for years. It was a small company that did not give much attention to publicly announcing titles when people were hired, but I did find out that he was brought in at a director level with comparable pay, even though he was doing the same job as me and another woman—and we'd both been working there longer than he had. It could be that he negotiated better than we did and had more years of experience outside of the company, but in that company, his work was at the same level as ours. Yet he was given at least twice the title and salary, and of course he had the attitude that he was above us, even though none of us were told his role was anything but our same level.

Women often feel like they are singled out and their work is more heavily scrutinized, and they are punished more harshly for any mistakes. A mistake by a woman is called incompetence, while if a male makes the same mistake, it is called an accident or a fluke. Audrey says:

Women are judged harder than men. Your work is scrutinized a lot closer; you have to be more perfect. That's one thing I learned: you have to be perfect—imperfections are not tolerated. I remember a time that a male colleague was supposed to be doing backups on a Unix server tower. Apparently, he had been having three weeks of errors on the backups but didn't tell anyone and went on vacation. The servers crashed while he was gone. I had to restore all the way back to before the errors and apply all the updates to fix the problem. When he got back, he was given a promotion even though I fixed the problem. Mistakes are overlooked as a man; if I had been the one to mess up the backups, I would have been fired.

And women judge each other just as harshly or even more so than men.

When a woman speaks out or complains about her treatment, she may be labeled "too aggressive" or a complainer.[22] The author of the book *Crucial Conversations* explains it this way:

An emotion-inequality effect punishes women more than men. Women are burdened with the assumption that they will conform to cultural stereotypes that typecast women as caring and nurturing. Speaking forcefully violates these cultural norms, and women are judged more harshly than men for the same degree of assertiveness.

I had a situation where a colleague was very demeaning to me when I noticed a problem while testing some code. I was told that I did not know what I was talking about and had looked at it wrong. When I discussed it with his manager, I was told he was just that way and that I should accept it. Again, I asked if that male manager would tolerate that behavior toward them and got uncomfortable silence, but the situation was later quietly resolved. It did not hurt that I was right about the error and my proof was sound and thorough!

Even female experts in male-dominated fields tend to contribute less to group conversations and thus are less recognized than male experts. Women with expertise in the subject are questioned more when they offer their input, if they are even allowed to speak as an expert

22 Liza Mundy. "WHY IS SILICON VALLEY SO AWFUL TO WOMEN? (Cover Story)." *Atlantic* 319, no. 3 (April 2017): 60–73.

and not assumed to be a junior member. According to Lisa, who has worked in insurance and finance in typically male-dominated departments,

Women need to have more education and credentials to get the jobs with titles and pay. Women are constantly needing to explain or display why they are good for the job; why they made a decision; or to be heard. If we fail, the opportunity is quickly turned over to a stronger individual instead of being mentored. I've learned to adapt by becoming an expert in my skills. The challenge is I intimidate some male/female managers, who will then set out to make waves.

It's a double whammy for women, as they say males are judged by their potential where females are judged by their past performance—which has already been harshly judged.

Another memory from my career: I was told that I was being overly emotional or aggressive in a group discussion when I was simply trying to contribute information that I felt was being ignored. I was also told that I should try to be nicer. I point-blank asked my supervisor if there was anything I said that was incorrect, and asked if they would say the same thing to a male in my position, and I got uncomfortable silence. But when women do not speak up and are not recognized for their expertise, they are passed over for leadership roles and advancement.

Along with not feeling like they belong in certain male-dominated fields, women often feel a sense of shame for not doing what they are supposed to—not acting as society feels they should as a woman. It is no wonder girls and women hear these struggles and choose not to even try to enter these fields. When you know going in that the game will not be fair, and you will be held to higher expectations just based on your gender and scrutinized heavier, it can be hard to convince some that the payoff is worth it. However, we know these fields are our future and will have mass shortages. So, maybe we need a "Rosie the Riveter" type of campaign showing women as the heroes who come in and save the field!

Chapter 5

WOMEN IN LEADERSHIP

As women progress in the field, they may want to move into a leadership role, but they face stereotypes that women do not want to be or are not suitable to be leaders. Many believe that leaders are born with certain genetic qualities and characteristics that make them desire to be leaders—traits such as drive, intelligence, and competitiveness, which are all stereotypically considered masculine traits. As previously discussed, girls are told early not to be too bossy or too outspoken, and this expectation carries into adulthood. Adding to this, women tend to get pigeon-holed into roles around the stereotypical strengths of administration or relationships, roles that are more teamwork-oriented or customer-facing, and less in areas that lead to leadership promotions.

As a result, women lose their voice in leadership.

These stereotypical expectations lead to people expecting female leaders to fail. Even if a woman does comparable or exceptional work compared to her male colleagues, her work is held to a higher standard and evaluated more harshly,[23] or she has her work devalued or ignored. As Audrey, who worked in IT for over thirty years recounts:

When I was up for promotion (to a high-level technical IT position), there were very few spots available. I was told I needed to know a certain application to get promoted, which I didn't. I asked all of the men who got promoted that year if they knew that application. They did not. The company gave me a raise money-wise, but since I didn't have the title, I wasn't eligible for the bonus and stock perks that go with the title."

Katherine Coffman has found that gender stereotypes even cause women to degrade their skills. When given positive feedback, "WOMEN ARE MORE LIKELY THAN MEN TO SHRUG OFF THE PRAISE AND LOWBALL THEIR OWN ABILITIES."[24] Yes, I am shouting that quote. Why do we do that?

We tend to attribute success to the team rather than recognize our own talents because we are trained from a young age not to talk about ourselves and told not to brag. We have a tough time discussing our accomplishments rather than how the team succeeded. This is great for team building and working with others, but it backfires when people are viewing women in their individual roles. We lessen our part in the team. This often comes out as a woman saying, "I'm only a ..." or "I just..." instead of stating her role or position. I have caught myself

23 Holmberg-Wright & Wright, *Education Innovation Journal*, 51-58.

24 Gerdemen, Dina. (2019). "How Gender Stereotypes Kill a Woman's Self-Confidence." *Working Knowledge: Business Research for Business Leaders, Harvard Business School.* Web. https://hbswk.hbs.edu/item/how-gender-stereotypes-less-than-br-greater-than-kill-a-woman-s-less-than-br-greater-than-self-confidence

doing this, downplaying the importance that I bring to the table as a Project Manager. I will joke that I don't do any "real" work. What I mean is that the team is doing the coding or configuration and that I am organizing it, but I downplay my contribution to keeping everything on track, communicating, and removing impediments.

As a society, we still hold the view of authority figures as male, so when a woman gets into one of these roles, she faces a no-win situation. If women are too stereotypically female, they may be considered to have a "lack of fit" for advanced roles, but if they act too stereotypically male, they are trying too hard. Her upward mobility and successes that she achieves may also be attributed to having masculine traits rather than skills and competence. When women do attain roles such as executive management, if they are aggressive and emotionally tough, they are often viewed and evaluated less favorably by both men and women because they are not sticking to the gender stereotypes. People like them less.

Contradictorily, if a woman behaves to her gender norms by being kinder and concerned with the team, she is seen as too weak for the executive roles.[25] She needs to behave like a man, but still behave like a woman! Jackie has seen this in action:

I have found that it takes longer to progress to senior positions unless you play the game well and make sure you don't come across as too domineering. Equally, I'm not one to hold my tongue so I have had to find acceptable (to others) ways to express my opinion. I've not come across a

25 Heilman, Madeline E., Aaron S. Wallen, Daniella Fuchs, and Melinda M. Tamkins. "Penalties for Success: Reactions to Women Who Succeed at Male Gender-Typed Tasks." *Journal of Applied Psychology* 89, no. 3 (June 1, 2004).

single male who has ever had to make these adjustments to their professional conduct.

Conversely, negative experiences which would typically be considered an "off day" for a man may be seen as a failure for a woman.[26] After a particularly tough meeting, Jackie recalls the following discussion with her manager: "I was told that there were appropriate channels of communication in the organization and that I should consider my commitment to the organization when challenging his direction. Male colleagues who said just as much or more in meetings were never spoken to." Even when the woman is seen as competent as a leader, she is often shunned by their peers and feels isolated and disliked, and she may feel she needs to decrease or downplay her leadership role to be liked.[27, 28]

As stated previously, stereotypes are prevalent because they are easy ways to categorize others, and people use them to dismiss others regardless of knowledge, skill, or talent. We make assumptions based on stereotypes about how a person should behave, look, or feel. The problem is, even if there is truth to the stereotype, stereotypes cater to the average. If all women want to be is average, then we can fall easily

26 "Discounting Their Own Success: A Case for the Role of Implicit Stereotypic Attribution Bias in Women's STEM Outcomes." *Psychological Inquiry* 22, no. 4 (October 1, 2011): 291–95. doi:10.1080/1047840X.2011.624979.

27 Trauth, Eileen M., and Debra Howcroft. "Critical Empirical Research in IS: An Example of Gender and the IT Workforce." *Information Technology & People* 19, no. 3 (August 2006): 272–92.

28 Koenig, A. M., Mitchell, A. A., Eagly, A. H., & Ristikari, T. (2011). "Are leader stereotypes masculine? A meta-analysis of three research paradigms." *Psychological Bulletin, 4.* Web. https://doi.org/10.1037/a0023557

into the stereotypes, but I think women are and want more. As with any average, the only way to change it is to skew the numbers in one direction or another. This is not to say all women need to go take testosterone and become men. It might mean creative ways to use tools or knowledge—and that benefits everyone. Isn't that a win for humanity?

The more women are viewed as capable individuals in their own right, the faster the stereotype will change or disappear. If a girl or woman shows ambition and drive to accomplish any dream she has, it shouldn't be considered an anomaly, that she's driven "for a girl." If a woman wants to take a leadership role, she should be evaluated on her skill and talent against other candidates, not how much she behaves as a woman should, or like a man. Everyone brings different styles to leadership, and each works in different situations, and many of the stereotypical masculine traits such as aggressiveness are not always seen as positive in leaders in today's world. Having more women leaders means the stereotypes about what makes a good leader will change, and whether men or women are better leaders will change too, opening doors for all to be judged equally.

Chapter 6

FIGHTING STEREOTYPES

What can we as individual women do to fight back against bucketing? How do we say that the bucket doesn't fit or isn't even the right bucket? How do we get rid of buckets completely or create an understanding that the buckets are too simple and broad? The first thing to do is to be as knowledgeable as possible about all aspects of the job and the typical stereotypes so you are prepared to prove them wrong. Never stop learning about new trends and what to expect in your field. Be prepared for meetings or discussion topics. Be ready to speak up and give input.

If you are in a male-dominated field, be prepared to work just as hard as the men or even harder. And if that means physical labor, then make sure you are ready for the work, or know and use ways to make the workload easier. You might even get some ideas implemented to help all of your coworkers! If you do not need help, do not let a male assume they need to help you. Kindly thank them, but let them know firmly that you can handle the task. If you need help, ask specifically for what you need help on, and do not let anyone take over the whole task.

If you hear anyone say that you cannot do something because you are a woman, or that you did something well for a woman, ask them to clarify what they said. Ask them what being a woman (or any characteristic, for that matter) makes it impossible or surprising that you can do the task. Politely confronting people about their beliefs in stereotypes and getting them to think through what they say will engage their brains and change some of those long-held beliefs. It could be that they simply are parroting what they have always been told or heard, or have never seen a woman do a particular job. A quick judgment based on stereotypes! If we can get them to see the individual and consider what they believe, we might just change some minds and remove some phrases from their vocabulary.

If there is a role, project, or task that will stretch your learning and abilities, go for it! Even if you think you do not have the skills or knowledge, the best way to learn is by doing. Find out what you need to be successful and if you need additional education or training, ask for it. If the company will not provide it, look for outside sources. There are many free ways to find information through the internet and free or affordable courses through places such as Coursera. Check with industry organizations too; they may offer free or inexpensive training or resources. Use your network to find someone with that education or skill, and ask if they can point you to resources or help you gain knowledge. The key is not to just assume you cannot do it and give up before you try. Always put your hand up and show up as the person right there, willing to try anything that a male will. You will gain respect from your coworkers and supervisors, learn new skills to leverage for promotions, and show that women can do anything just as well as men. Do not downplay your talents or what you learn. Make sure people know that you were part of that project, that team, that success!

If you are told that you are too direct or abrasive, consider the source first. Is this a manager or other leader offering coaching advice and constructive feedback, or a colleague who is uncomfortable with a strong, outspoken woman? Sometimes, we say things that are taken wrong or that we do not mean. Learning to be concise and use the right words is part of learning good business communication. So, did your words or tone unintentionally (or intentionally!) hurt someone's feelings or convey the wrong message? Ask the person to clarify what they mean and give specific examples. If they offer specific feedback, then consider what they are saying and watch how other people (men) speak in the same situation.

This does not mean you have to sugarcoat your language. If you are being direct but not hurtful or defensive, and your messages are clear, then do not change your style. If you are using the same communication style as men in your role and organization, then do not change just because you are a woman. Evaluate the feedback closely. If you are going to take the advice, understand what needs to shift and why. If you decide not to take the advice, at least you have thought it through and know why you are not taking that advice.

If you need to give feedback or help out another woman, make sure you have very specific examples and consider whether you would penalize a man for saying the same thing. I recall an example where a young outspoken woman called out a senior Vice President in a large meeting for not responding to her email inquiries. We were all aghast. After the meeting, I took her aside and explained that a large meeting like that was not the forum at all for that type of discussion. I appreciated her directly asking the VP why she had not responded, but she needed to be mindful of the audience. This was not because of her gender; it was a mistake that she hopefully learned from, as would any team member male or female.

In my case, if I have been accused of being abrasive, I have specifically considered the feedback. In one case, I had been defensive unnecessarily, as I was in a new job after leaving a very difficult and harsh team. I was holding onto habits from that experience, which was not the fault of the new team, and taking it out on them. They appreciated my directness but not my tone or style. The feedback was correct and helped me to realize what I was doing and change my approach—not my directness but my tone and assumptions before the conversation started. Once I did that, the team worked with me better and never complained about my communication style though I was the only woman.

The same consideration for feedback applies in other areas such as dress. If someone is looking or saying something about what you are wearing, look around and consider what everyone else is wearing. If the men are dressed in jeans and t-shirts and this is acceptable and you dress the same, then point out that you are wearing the same thing as everyone else. It is okay to have your own style, but you should fit into the corporate culture. If the culture is to wear professional suits and be very dressy, you should try to match as much as possible. You can have your own style with accessories or colors, but do not dress like you are heading to the club if everyone else is dressed more somberly. You may have to save some of your personal style for off-work hours and just consider your work apparel as a uniform.

If you are looking to get promoted, the saying is to dress for the job you want—so look at how your superiors dress. If they are clothed a little better than their subordinates, try to match that level of apparel. This does not mean we have to show up in skirts and high heels. I hate wearing skirts or dresses in the workplace but have found that if I need to be "dressy," nice slacks and a pretty blouse or polo are both comfortable and fit in just fine with the males in my industry. Pay attention to what others wear and find what works for you.

Most of the stereotypes women face just need to be proven wrong by those in the field. As we encourage more women to join male-dominated fields, the perception will shift; it just takes time. In the meantime, women need to not be afraid to do these jobs. We need speak up for ourselves and address any discriminatory behavior, even if covert, when appropriate. And we should encourage and support other women in our field, as well as those wanting to join it. We have to be willing to say "that's not true" and prove it.

To summarize:

- Step up and show your talents to the world.

- Take risks. Push yourself to try new roles, projects and tasks. Ask for it!

- Participate in discussions. Speak up when you have information. Offer suggestions for change that will help everyone.

- Consider negative feedback carefully. Who says it? What is the context? Does it have merit? Would the same be said to a man in the same position?

- Be an individual within the guidelines of business norms for your industry.

- Speak up against stereotypes, and help other women facing those stereotypes.

- Challenge people to examine their beliefs in stereotypes by asking directly why they believe them.

Chapter 7

UNCONSCIOUS BIAS

Not only do women face outright stereotypes that tell you how you should look, act and feel; we also face unconscious biases which are an extension of belief in the stereotypes. Unconscious biases are those that we may not even realize we have toward others. They have been built as a result of hearing certain messages over a lifetime from family, our culture, and our media, and they are ingrained in our subconscious. Gender biases are "the powerful yet often invisible barriers to women's advancement that arise from cultural beliefs about gender, as well as workplace structures, practices, and patterns of interaction that inadvertently favor men" according to a 2018 article by Madsen & Andrade.[29]

Even when people reject outright gender stereotypes, they are still likely to hold unconscious bias because they are hidden in messages and behaviors. The types of bias that women in STEM face include the view that if she is too feminine, she is not seen as competent, but if too

---

29 Madsen, Susan R., and Maureen S. Andrade. "Unconscious Gender Bias: Implications for Women's Leadership Development." *Journal of Leadership Studies* 12, no. 1 (Spring 2018): 62–67. doi:10.1002/jls.21566.

masculine she is not likable. People unconsciously expect men to be in charge and will turn to a male in a conversation, thinking they automatically are the leader. They will unconsciously leave women out of communications, which means women miss important information, input to decisions, and knowledge sharing. A woman with children or of child-bearing age may not be seen as committed to the job because of family duties. There are age biases as well—if too young, she is seen as less capable, but if too old, she's stuck in her ways.

Bias also shows a tendency to gravitate toward people who are similar to yourself, known as affinity bias. In male-dominated fields, men are used to working with other men and may be uncomfortable working with women. Hiring managers will hire people they are familiar and comfortable with, so a young male manager will tend to unconsciously hire young male employees. Even a hiring panel, if not staffed with a diverse set of managers, will tend to hire similar people due to affinity bias or even conformity bias by agreeing with what others on the panel say. They often just go with what the group says, even if there is a better candidate. This problem perpetuates the biases as teams are built with similar people, and it is tough to break through that barrier if you are different— whether that's gender, race, or any other characteristic or attribute that can separate someone. If a person does not see others like themselves in any situation, it can be daunting to think of breaking through to join that group as an "only one."

The stereotype of how a woman should dress often shows up as beauty bias. People tend to give more favorable treatment and higher opinions to those they consider more attractive. This can lead to people being hired for their looks rather than their skills, or getting more attention for projects because they are more attractive. A friend used this to her advantage to get upgraded equipment for her team: "A pair of red heels and a low-cut red blouse was all it took." Sadly, this was after presenting the reasons the equipment was needed and getting turned down. Conversely, a person could be overlooked because they do not dress the part, as a woman who does not dress conservatively enough may be thought of as less intelligent and therefore not suited for the STEM fields. Beauty bias leads to age bias as our looks change as we age. Older women are not considered as pretty and likable.

Bias affects behavior expectations too. If a woman hears that she's "too direct," this is inherent bias because it's not how women are expected to act. Leaders will even discipline a woman who is too direct for having a harsh tone. It is assumed that women will be soft-spoken, meek and timid, and hold back from speaking up. She will use language that is softer and dance around the hard topic to make sure everyone feels okay with what is being discussed. Women do tend to use language that is more inclusive to a group, but this makes them seem less confident and knowledgeable in the topic. Even replacing an "I" in a sentence with "we" to be more inclusive lessens the value given to the individual's contribution to the topic. Women do have a slight advantage with language and language processing, as shown in brain studies, and they use this to help garner group cooperation when needed, but this is not proof that any woman is not able or does not desire to use direct language. And when a woman does use direct language, she hears negative feedback about it. And it's often other women saying, "I can't *believe* she just said that!"

We also tend to use a higher pitch to soften our asks. If you have ever heard a woman use the higher-pitched voice with "Could you please..." it is an attempt to soften the request. A male can tell a subordinate to do a task, while women think we have to ask as if it is a favor from the other person. We also tend to apologize for things we are not responsible for, sometimes in an effort just to keep the peace or get the other person to apologize. It is annoying and makes people look at you as less of a leader when you constantly use the falsetto voice and apologize for everything, but you can also be seen as too forceful if you adopt the authoritarian voice and tell people like a male would. So for women, it is a balancing act to find the right voice to use.

And it's not just the words used; it's also tone. We have an inherent bias to hear higher-pitched voices as shrill or gossipy. Women with higher-pitched voices are said to have baby voices, which supports the bias that a woman will be soft-spoken and meek. According to Laurie who has worked in radio for over twenty years, audiences "love a 'sexy' voice" in women. At the same time, she notes, "When I first started in radio, we were told to never play two female artists back to back, and to play at least four males in between. Because their voices sounded too much alike... As far as I can tell (from) listening to the radio, that rule is still in play."

Women who give speeches specifically train to keep their tones and emotions in check so as not to be described as shrill or emotionally out of control. Women in politics and speaking are even told to try to use lower pitches to come across as more authoritative and credible. There is some truth to the fact that lower voices and deeper pitches carry further and tend to be better heard, so women can consider that when making speeches and presentations. But it seems ridiculous that vocal cords and air passage through them determine someone's credibility, status, or knowledge. Yet that is how inherent bias works; we believe it in our core without thinking it through. This is nothing new either. Susan B. Anthony once complained, "No advanced step taken by women has been so bitterly contested as that of speaking in public. For nothing which they have attempted, not even to secure the suffrage, have they been so abused, condemned, and antagonized."

Even hand gestures and overly showing emotion during speeches lends to bias against women, and she is called hysterical. The bias toward women demands that we speak calmly and politely, which unfortunately also leaves us open to being talked over. If a woman even in the highest offices speaks too politely, it is seen as a chance to interrupt her and dismiss her message.[30] Women who speak up "too much" have their words devalued, and if a woman displays emotion such as anger, she becomes known as an angry woman rather than a woman reacting to a situation or circumstance.[31] Facebook COO Sheryl Sandberg refers to this as "speaking while female." In male-dominated organizations, women are often interrupted or talked over when trying to speak, or their ideas will not be heard until a male repeats the same

---

30 Tonja Jacobi and Dylan Schweers. "Female Supreme Court Justices Are Interrupted More by Male Justices and Advocates." *Harvard Business Review*, 11 April 2017. Web. https://hbr.org/2017/04/female-supreme-court-justices-are-interrupted-more-by-male-justices-and-advocates

31 Victoria L. Brescoll and Eric Luis Uhlmann. "Can an Angry Woman Get Ahead?: Status Conferral, Gender, and Expression of Emotion in the Workplace." *Psychological Science.* 008 Mar;19(3):268-75. https://doi.org/10.1111/j.1467-9280.2008.02079.x

idea and gets the credit. This means many women just give up and do not speak out in groups.[32]

If a woman is showing assertiveness in front of others, we are biased to think of her as being a bitch who is out of line. This isn't a man's view only; women have similar inherent biases and expectations of how a woman should act and speak and will gossip about others who are not fitting their view of the gender. Women who display too much emotion get negative reactions and lose status in people's minds, even being seen as less competent in professional situations.[33] Tina says of women in IT: "I think we are looked at like heartless, bitchy, overly opinionated women. When in reality, we are just motivated, educated women with a lot to offer to our employers. We just have to work twice as hard to be seen and heard."

This makes sense. Women are "supposed" to be the great caregivers and nurturers of the world and allow all voices to speak and be heard, while men are "supposed" to be more assertive and in control, which historically has been supported by the proportion of men in leadership roles versus women. Women are often asked to come into a team situation because they are so good at getting everyone to get along. Or they are asked to handle all communication because women communicate better. These are unconscious biases that place women in the areas of nurturing others, whether the woman is suited

32 Sheryl Sandberg and Adam Grant. "Speaking while Female." Opinion, *The New York Times.* 12 Jan. 2015. Web. https://www.nytimes.com/2015/01/11/opinion/sunday/speaking-while-female.html

33 Brescoll, V. L., & Uhlmann, E. L. (2008). "Can an angry woman get ahead? Status conferral, gender, and expression of emotion in the workplace." *Psychological Science*, 19(3), 268–275. https://doi.org/10.1111/j.1467-9280.2008.02079.x

or not. These attitudes and their results also reduce the chance for others to step up and learn those roles and develop team-building and communication skills.

Chapter 8

FIGHTING BIAS

How do we fight unconscious bias when people do not realize they are being biased or understand where their opinions even originated? Let's acknowledge and realize our own biases to start. We need to check how we think about women that we meet. Do we judge her on her looks first? Do we make assumptions about her based on her gender or mannerisms, or do we listen to what she has to say and offer? Are we falling into perception bias by deciding another person's actions or behaviors simply because of their gender? Do we make assumptions about roles based on gender without asking?

Remember, if someone says or does something that is showing an unconscious bias, ask them, "What do you mean by that?" This simple phrase gives the other person a chance to walk back what they said and think about what they meant without forcing an angry confrontation.

Both men and women have unconscious biases, so programs to point these out can be helpful to mixed audiences. We need to be able to have open and honest conversations about what women face in the workplace, even if overt discrimination no longer happens. Talking about situations where bias has happened will help bring light to these situations, and invite everyone to consider how people interact with each other and the effects our interactions have. We are going to have to be candid about situations that we all have been in, where we let biases drive our words and actions. We must change ourselves and teach our children not to make assumptions about anyone based on what they see. Rather, we can teach them to be curious and ask questions first.

As for helping women in the workplace, hiring and promotions need to be decided by a diverse panel of people to avoid any gender or affinity bias. As much as possible, the process needs to be made neutral, even if that means stripping out names in resumes to focus on just the person's skills. This will help with anchor bias too, where one piece of information such as a feminine name will cloud a person's judgment about the rest of that resume.

It will take time to change everyone's mind, and someday hopefully we can get to a place where people look at each other only on their own merits, rather than preconceived notions of who they are based on gender, race, or any other factor. Until then, we cannot ignore the fact that we need to overcome these biases within people, even if they say on the surface that they are not biased because they do not even recognize it within themselves.

To summarize:

- Watch for and challenge unconscious bias from any gender.

- Speak up if someone says a woman is being overly emotional or too aggressive.

- Listen for phrases in our speaking that may indicate an unconscious bias and eliminate them.

Chapter 9

JUDGMENT

Stereotypes and biases describe how we expect people to act, and when they act outside of these roles, they face harsh judgment from others. For women, this criticism seems to come most often from other women. "If I have been judged crucially, it was by other women. Men who have been my superior have only provided praise. Mainly because my role in the organization either makes or breaks my superior's position. I have always delivered. I have always set them up to succeed," says Sarah about the healthcare field. And Jennifer says of retail that judgment has not gotten better even though the field is dominated by women. "It's worse. It's more open and blunt. It's not the men judging. It's the women. Women do not treat each other nicely. I'd rather work with men!"

Women and men both see an assertive woman as a "ball-buster," or overly aggressive toward men, or more plainly, a "bitch." Men not displaying masculine traits are called "girly" in a condescending way, as being labeled female is considered to make you "less than." These derogatory types of words hold power to lower women's status or put her "in her place." A female lawyer who has worked in various areas of law for eighteen years recounted to me:

There are a lot of powerful women figures in the legal field. I think the tendency is that women are 'bitches,' and many are very intense. Most of the women I know are great people, but we need to take strong positions to advocate for our clients as our job requires. Unfortunately, that does impact how people view women in the legal field.

Outspoken women are subjected to public criticism, hurtful comments, or even silent but still hurtful eye rolls and shunning.

We have been conditioned to be competitive against other women to keep ourselves up front and on top in everyone's eyes. Women often feel that due to limited advancement opportunities, they need to tear each other down to make themselves appear better. They may hang around the boss more trying to gain favor or attention, or find ways to minimize other women's accomplishments. They spread rumors that she slept her way to the top, as if sex is the only way a woman can obtain power in a patriarchal society. Or that she is a token member of the organization so they can show they have at least one woman and therefore are diverse. These hurtful thoughts and comments will show through in interactions with each other, making her feel cast out and alone at a time that she needs the most support from other females to navigate and survive the male-dominated system.

Women who want to work or advance in their jobs face judgments about taking time away from family, even though not all circumstances allow for women to stay home with the children. The stereotype of women being the caretaker, homemaker, and child-rearer means that a woman who decides to break from those roles faces judgment as a bad mom and wife. If she leaves her children in daycare or with a nanny, she is asked why she wants others to raise her children. A father who takes care of the children so the woman can work or participate in activities outside the home is "babysitting," while it is simply expected that a woman gives up her time to care for the children if the father has outside obligations.

If a child is sick, women are expected to be the caregiver, and she faces judgment if she does not take time off to stay home with the child. As related by Tina, an IT professional with children, other women thought "that I should miss work because one of my kids was sick. I had it covered. My spouse was a stay-at-home at the time and would cover any kid issues. I was looked at as an uncaring mother. Almost insensitive on some level because I should be home with my kid." Jennifer, who has worked in retail for decades, has faced this same judgment in her working life: "They assume it is always going to be the woman taking time off for family reasons (sick kids). They don't think of the dads taking time off." A busy woman who does not have time or does not want to participate in things such as volunteering in a child's classroom or who misses a child's activities due to other commitments is considered selfish. Women who work long hours are not meeting their wifely or motherly duties, even if the woman wants both a career and family and is handling it all successfully. Working mother guilt is common.

Women who do not want or cannot have children are questioned why and told how much they are missing out of life by other women. Paula, a woman who has been in social work for thirty-seven years, has faced trouble due to not having children: "I don't have kids and for some clients, there's been a belief that I can't help them because I don't know what it's like to be a mom." This happens even though she has worked in the field for over thirty years with many different patients and families. If a woman is waiting to have children until she is older so she can further her career, people wonder why bother, because it is assumed she will quit once babies come along. Or, she faces the judgment against working mothers. A woman is told having children will wreck her career because she will be distracted from work. It's believed her performance will suffer because she will need to take time away to care for the child, and not put in the effort sometimes required due to home commitments. But if she does not want children, then she is judged as not acting as a woman should by bearing children. And as Jennifer has found when trying to reenter the workplace after raising children, people "assume there is something wrong with you if you take off time to raise a family and try to reenter the workforce when they are teenagers. It is considered 'late'—sometimes 'too late' to start a career."

Related to the stereotype of women having a better appearance, women are judged on competence, warmth, and likability based on their looks, much more harshly than men. Political figures, newscasters, and celebrity women's appearances are often discussed and used to try to denigrate their competence, as shown in the political campaigns of both Sarah Palin and Hillary Clinton. Women are taught and culturalized from a young age that beauty makes you a better person. Even as children, girls are judged more by their clothing than boys in terms of intelligence and academic potential.[34] Later in life, women tend to be harsher judges of what other women wear, how they style their hair, even how they wear their makeup in comparison to the beauty ideal as a way to boost their own self-esteem.[35] However, a woman wearing too much makeup or dressed too sexy is seen as less competent too! "Women are aware of the way they are perceived when they emphasise appearance over more competence-based qualities and this provokes behaviours such as distancing themselves from other sexualised women that portray themselves in this light." The higher the woman's position, the more negatively she is viewed by other women if she dresses overly provocatively—that is, if she wears a slightly shorter skirt or unbuttons too many buttons on her shirt. "Even minor portrayals of female sexuality in clothing can have a negative impact on the judgments made by females of a female senior manager."[36]

Paula, the social worker, has faced judgment on her looks too. She explains:

Therapists can alienate themselves from one another because there is often a perceived need to not show any 'flaws', not to admit to any strug-

34 Behling, D.U., & Williams, E.A. (1991). "Influence of Dress on Perception of Intelligence and Expectations of Scholastic Achievement." *Clothing and Textiles Research Journal*, 9, pp. 1-7.

35 Nathan A.Heflick, Jamie L. Goldenberg, Douglas P. Cooper, Elisa Puviab. (2011, May). "From women to objects: Appearance focus, target gender, and perceptions of warmth, morality and competence." *Journal of Experimental Social Psychology*, 47(3), pp 572-581.

36 Howlett, N., Pine, K., Cahill, N., Orakçıoğlu, İ., & Fletcher, B. (2015). "Unbuttoned: The Interaction Between Provocativeness of Female Work Attire and Occupational Status." *Sex Roles*, 72(3–4), pp. 105–116. https://doi.org/10.1007/s11199-015-0450-8

gles. I'm overweight, which is a 'flaw' a person can't hide. I think there have been times when I wouldn't get certain clients referred to me because of my weight.

Obese women are judged more harshly than even obese men by hiring managers, and are less likely to be considered for leadership potential and promotions. They are seen as sloppy, lazy and less intelligent purely based on body size. Some studies have shown that heavier women are sixteen times more likely to experience weight discrimination than males in the workplace. Overweight men in politics actually received higher ratings than thinner men, but overweight women in politics receive lower ratings across the board.[37]

It has been said that the enemy of good is perfect. Women get in their own way trying to be perfect. We think that we have to be ideal mothers if we have chosen to have children. We compare ourselves to other mothers and what we see via media and feel like failures because we are not perfect. We judge ourselves against working mothers and feel that we fall short of being able to succeed at work, and we fall short when we judge ourselves against mothers who have chosen to be home and focus on their kids instead. We look in the mirror and feel ugly or fat or not stylish enough. We let these feelings bleed over into everything. Suddenly, we are not good enough to have that corner office or get the job that we want. We are judging ourselves against each other and an ideal of perfection that does not exist. No wonder we cannot see our strengths and skills and what we bring to the workplace.

---

37 Kelly King, MPH, and Rebecca Puhl, PhD. "Weight Bias: Does it Affect Men and Women Differently?" Obesity.org. Spring 2013. Web. https://www.obesityaction. org/community/article-library/weight-bias-does-it-affect-men-and-women-differently/

Trying for perfection is exhausting. Perfectionism has been tied to depression, suicide, eating disorders, anxiety, obsessive-compulsive disorders, and insomnia. We are so anxiety-ridden about doing everything exactly right that it pervades our whole being. We are so overly critical about ourselves. We cannot handle making a mistake, or not being as good as someone else, and downplay our own performance and achievements as a result.[38] When we fall short of our overly high ideals, our self-esteem takes a huge blow and we give up. We believe we just are not good enough. We have judged ourselves as not worthy before we even really have a chance to try anything. We defeat ourselves.

Why do we strive to be perfect? Studies show that perfectionism tends to develop during adolescence and is driven by the parent's level of expectation and belief in perfection.[39] Many women can remember middle school being a time when other girls started to become particularly nasty toward each other, judging each other on their looks, style of clothes, even on their friends and social circles. Girls are pushed to be perfect by perfectionist parents and are also getting cues from their peers that they are not perfect, and thus the judgment spiral begins. We are teaching our children negative, harmful patterns by trying to make them perfect.

Women are aggressive toward other women in the workplace to put themselves at an advantage for projects and positions. This

38 Masson AM, Cadot M, Ansseau M. [Failure effects and gender differences in perfectionism] L'encephale. 2003 Mar-Apr;29(2):125-135.
39 Lavinia E. Damian, Joachim Stoeber, Oana Negru, Adriana Băban (2013, Oct). "On the development of perfectionism in adolescence: Perceived parental expectations predict longitudinal increases in socially prescribed perfectionism." *Science Direct*, 55(6), p 688-693.

may hearken back to our roots in history, wanting to attract the best mate and competing for scarce resources. Since direct aggression is frowned upon in females in the workplace (and in general life), women use indirect aggression—or ways to make other women look bad by spreading gossip or rejecting them from a social group.[40] Girls start this behavior at a young age, and as girls age into women, they find ways to better hide their aggression to reduce any retaliation and reduce any judgment on themselves.[41] These indirect attacks can completely ostracize the victim from certain groups and social circles. This hurts the esteem of the target, leading to loneliness, depression, and anxiety among these young women which can be long-lasting in its effects. Unfortunately, this behavior carries through to adult life and the workplace.

While men tend to be more aggressive and direct, confronting their perceived adversary face to face, women are sneaky and try to be indirectly aggressive while staying anonymous. Women show this indirect aggression in the workplace by spreading lies about another woman's skills or how she advanced, or otherwise downplaying her achievements. They may spread rumors that she slept her way to the top or was chosen for her looks over her skills. Many women tear each other down to raise their own worth, but try to hide it so that they cannot be later blamed. Often, a female will be "nice" to another female

---

40 Richardson, D. R., & Green, L. R. (1999). "Social sanction and threat explanations of gender effects on direct and indirect aggression." *Aggressive Behavior*, 25(6), 425–434. Web. https://doi.org/10.1002/(SICI)1098-2337(1999)25:6<425::AID-AB3>3.0.CO;2-W

41 Björkqvist, K., Österman, K., & Lagerspetz, M. J. (1994). "Sex Differences in Covert Aggression Among Adults." *Aggressive Behavior*, 20(1), 27–33. Web. https://doi.org/10.1002/1098-2337(1994)20:1<27::AID-AB2480200105>3.0.CO;2-Q

face-to-face, while spreading negative gossip behind her back. Tina relates a story about another woman in IT who felt threatened by her and waited for negatives:

She went out of her way to try and find mistakes I made so she could point them out to the boss. It was the most unpleasant work environment I have ever worked in. I ended up having a boss who watched me closely and a woman who stalked my every move. She would even listen to my phone calls.

Even with changes in the workplace to give more opportunities to women and highlight their skills and competencies, many still feel the need to compete with each other for attention and promotion.[42] Jennifer says she has learned to:

...keep your mouth shut about your education. Other ladies will become jealous. Their attitudes infiltrate and disrupt the workplace, which makes it difficult to advance. Who wants to put you in a management position when coworkers don't like you and choose not to get along with you? Twice, women have found out that I went back to school and felt threatened by me. It's better not to speak of your accomplishments, unless you don't care if you have friends and good relationships with coworkers.

Such women can become workplace bullies, or "Queen Bees." Queen Bees make the workplace difficult for other women. Whether they are directly, loudly aggressive or quietly passive-aggressive, they cut other women down to keep themselves in the top spot, instead of being allies to bring each other up. They might be called

---

42 Anne Campbell. (2004). "Female Competition: Causes, Constraints, Content, and Contexts." *The Journal of Sex Research*, 41(1), 16.

"catty" or "bitchy" by others, but the main effect is that women do not want to work for other women when they have experienced a Queen Bee. This behavior could be a consequence of the zero-sum game in male-dominated fields, that there are only so many top spots in general and even less for women, so some feel the need to compete and push other women down—and some do it in a nasty way. Or they feel they had to pay their dues to get where they are, so other women need to also pay.

A female lawyer has experienced this "right of passage" behavior, describing it like this:

There are times when women are harder on younger women because that is how it was for them when they entered the field. I believe that is changing slowly as more women enter the industry and break the glass ceiling. The firm I worked for in private practice was very good at giving women the opportunity, and that is one of the reasons why I chose to work for that firm. However, even being in a more 'friendly' environment, some women are penalized for their decisions to take care of children and the like and forgoing their careers.

There might even be a concern that women will be judged as biased when we help other women! When we decide and state that we do not like working for other women, we also feed into the biases and stereotypes that women are not good leaders.

Madeleine Albright, a former US Secretary of State, once said, "There is a special place in hell for women who don't help each other." Queen Bees not only fail to support and mentor their subordinate female colleagues, but they also demean and belittle the accomplishments of those who are trying or have advanced. The hell that women put each other through may be how that would look! These types of behaviors, exhibited by the Queen Bees, cost organizations in terms of losing good talent and knowledge, as well as lowering productivity and morale. By demoralizing other women, the Queen Bees cause them to lose interest in the job, doubt their own skills and competence, and ultimately leave the organization to get away from the abuse.[43]

To summarize:

- Remove judgment from ourselves about others' choices. Do not assume you know the whole story or have the best answer.

- Speak up against Queen Bees or others who use judgment and derision to put others below them.

---

43   Harvey, C. (2018). "When queen bees attack women stop advancing: Recognising and addressing female bullying in the workplace." *Development and Learning in Organizations, 32*(5), 1-4. Web. http://doi.org/10.1108/DLO-04-2018-0048

Chapter 10

OVERCOMING JUDGMENT

How do we change this culture of judgment? It may seem impossible since we are so conditioned to judge each other on looks, achievements, possessions, and even our desired lifestyle. In STEM, it should be easy to look past these things and simply focus on the person's skills and the attitude they bring to the table to solve problems. We need to be introspective and watch our first reaction for judgment. Are we really listening to what the other person is saying, or have we already made up our minds based on our judgment?

I think whatever a woman decides to pursue in life, she should have the backing of other women. If she decides to pursue a very technical or scientific path—great! She will bring great things to the field with her brainpower. If she decides to lead, then we support her ambitions. If she decides to quit her career and stay home to raise children, that's a tough decision too, but we know it is important to her to raise them to be our next great generation. If she decides not to have children yet, or at all, well, that's another tough decision too but we support her desires.

It's all about being supportive and not trying to force what we think on others. There are so many factors that go into every decision in life. We can't know what the other person is going through unless they tell us, and even then our view is based on our life and history. We need to stop judging others who choose a different path and support them in whatever they decide. If you do not agree or understand that path, try to learn more about it, or at least keep your judgments to yourself, support your fellow women, and call out others who don't.

As women, we often have a fear of being judged by others or worry too much about what others think about us. It can be difficult for women in the STEM fields to feel that they can stand up and show off their skills and abilities and face the opinions of others who do not think they even belong in the field, or who will criticize their work strongly just because they are a woman. Often, though, we are our own worst critics.

To start breaking free from this trap, women need to really think about why they feel inadequate or have negative thoughts. If this line of thinking stems from a place of fear of being judged, then we must work to overcome that fear, be proud of our accomplishments, and not be afraid to share them. When we catch ourselves mid-judgment, whether directed at ourselves or others, we need to stop, recognize the judgment, and redirect our thoughts.

Finally, people are going to judge you no matter what you do. As women in STEM fields, we have not only ourselves to consider, but future generations of women who need us to get past the fear of being judged and show the world what we can do. Let them judge. At least we are doing what we love and fulfilling our purpose.

Chapter 11

BENEVOLENT SEXISM

B enevolent sexism is a way that men will treat women as the weaker sex that needs protection. It is different from directly acting sexist because the person may feel they are doing it for the benefit of the woman. For example, they will not say that they do not think a woman should have a position, but instead, they treat the woman as a subordinate even when she is at their same level in order "to protect" the woman.

We have heard of "mansplaining," where a woman complains that a man will take something she has said or a topic she may already be an expert in and explain it back to her as if she is a child or lacks intelligence. In meetings, when a woman gives a presentation, a male colleague may feel they need to reiterate what she says or interject when not necessary. Males may step in and take over projects or assignments in an attempt to help, but really they just end up pushing the female away and not allowing her to learn and grow her skills. They may treat a woman as inferior in an attempt to help, such as grabbing boxes from her hands without being asked. Yet, I remember many times after a working lunch meeting where all the men got up and left and just assumed the women in the meeting would clean up!

These are not blatant sexist acts, and the male may not even realize they are doing it. But the effect is that the woman is put in a subordinate position to the male, even if they hold equal roles in the workplace. If she does not speak up or point out the behavior and lets the male continue, she misses chances to show off and advance her own skill and competency. Women may not recognize the effects and just think the male is being nice and will not think of or report it as sexism.[44] The behavior continues and women are still held back.

---

44 Hopkins-Doyle, A., Sutton, R. M., Douglas, K. M., & Calogero, R. M. (2019). "Flattering to deceive: Why people misunderstand benevolent sexism." *Journal of Personality and Social Psychology,* 116(2), 167–192. Web. https://doi.org/10.1037/pspa0000135.supp (Supplemental)

Chapter 12

MICROAGGRESSIONS

There are many forms of microaggression that both men and women engage in toward female colleagues. These behaviors include activities such as the previously mentioned case where a boss takes the males on the team out to a sporting event. They may not even do it consciously. If asked, they will report that women do not seem interested in those things. These men fail to recognize the importance of fraternity and networking that happens during those events, leaving women on the team even more outside.

Another microaggression is calling a woman names like "honey" or "sweetie" in the workplace. Some may think these are just nice words to use with women, but in fact, they are demeaning and have no place in the professional world. Telling a woman to smile more, or use softer tones instead of being direct, is another form of microaggression. And telling a woman that she is being overly sensitive to gender issues is a form of microaggression because it discounts her feelings and perceptions.

Women may not respond to microaggressions for several reasons. For some, these acts can seem trivial. Many women feel they will never be able to change others' behaviors. Or, she fears she will be seen as hostile or angry and cause workplace tension. Often, these microaggressions would be hard to prove as directly being gender-biased—though when I asked a male supervisor how they would feel if it happened to them, they did pause and reflect on the situation. It's hard when you are the only woman in the room and you feel like you have to fit in with the guys, or you are the model for not being "that" hysterical, oversensitive woman. How can we get common behaviors to change if we don't address them, though? Even small things build up over time and can make the workplace unbearable for women. And sometimes it might seem okay to continue or even act worse; give an inch and people take a mile.

When it is women who are aggressive or mean to other women in the field, we leave each other feeling isolated and alone, and often not wanting to work with other women. As Sarah relates from working in a mostly female-dominated field of healthcare, "I would rather work with men than women, to be honest. I don't like playing games, holding grudges or being catty. I am straightforward, often blunt, so I work better with men from that standpoint." If the attacks continue, the other woman may choose to leave rather than tolerate the behavior.

Women in management in particular are targeted for attacks, jealousy and social isolation. And this could be a reason that women either do not want to enter management positions or they leave—they cannot get support from either gender. They are not part of the boy's club, and the women treat them snidely. No one wants to feel shunned in their environment. This causes self-doubt and depression in the victim. If we want women to get into the advancement pipeline, we must find ways to keep good women from leaving the organization by recognizing and cutting out these aggressive and jealous behaviors.

Chapter 13

STICKY FLOORS, GLASS CEILINGS, AND GLASS CLIFFS

The effects of these stereotypes, biases, and judgments are the continued problem of women being stuck in lower economic and job levels. This hurts women looking to advance due to a lack of mobility or opportunities in the job market. The Covid pandemic in 2020 had a disproportionate impact on women in the workforce, even in the STEM fields, due to the closure of workplaces and daycares. Women in lower-level jobs found themselves without protection for leave or workplace flexibility if they needed to care for families that suddenly did not have activities outside the home such as school, child, or adult care, or if they or their family were exposed or became sick with the virus. Many of these jobs are perceived as ones that you have to do in person or are difficult to manage while taking care of family. Most of the telework options available are for higher-level positions in technology, management, and business functions, where we already know women are underrepresented. Lower-level positions are the hardest hit because of business closings as well.[45]

What is keeping women from the higher levels in STEM? As Leanin.org put it, women are not hitting a glass ceiling; we are facing a broken rung. What that means is that we are not getting promoted from those entry-level positions to that first step that will lead to future management or other higher-level roles.[46] Being stuck in these low-level positions means women miss out on educational and job training opportunities to increase skills and gain knowledge for the jobs of the future, leaving them in a poor position when organizations want to pivot and modernize, or needs to reduce workforce by getting rid of entry-level employees. It also leaves fewer women to promote from the ranks and less visibility for women who might make great managers and leaders.

We know that women are paid less than men on average, even if they have the same skills and are working the same job. This is often referred to as sticky floors, where women are mired down by lower-level jobs and pay scales, and cannot unstick themselves to move upwards. A woman who works in cybersecurity says,

I was the network team supervisor and learned that despite being graciously allowed on the team for diversity bragging rights, and despite

45 *Centering Equity in the Future-of-Work Conversation Is Critical for Women's Progress*. Center for American Progress. 24 July 2020. https://www.americanprogress.org/issues/women/reports/2020/07/24/488047/centering-equity-future-work-conversation-critical-womens-progress/
46 *Women in the Workplace: 2021.* McKinsey & Company. https://womenintheworkplace.com/Women_in_the_Workplace_2019.pdf

working as both a network engineer team lead and as the supervisor at the same time, I was being paid the second-lowest amount on the team and was making about half of what the top two team members were. There were no opportunities for growth that would be fair (since promotions were based on percentage increases), so I left for a position elsewhere in another industry.

Not only was the pay scale not fair; that company also lost good talent and domain knowledge— something that can take *years* to build. Even though she had proven her capability, she could not overcome her start on the sticky floor. It's a double whammy for women, as they say males are judged by their potential while females are judged by their past performance,which has already been harshly judged. Stuck to the floor by judgments!

We can also get stuck by being too good at the types of work these low-level positions require. If we are good at organizing people and events, we will be handed the administrative work. And we will do it, as women tend to want to please people and not say no. Males are less afraid to say no, or that they do not have time for less important administrative tasks. If there is the thought that no one else can fill that lower-level role, then the person doing those tasks either gets passed over for promotion, or maybe worse, has to do those tasks plus their new role.

And the more women try to advance, the more pushback they face in masculine-based workplaces. As a July 2017 now-famous Google employee memo starts, "I value diversity and inclusion, am not denying that sexism exists, and don't endorse using stereotypes." And then the email goes on to list various stereotypes on why women do not like or want to work in computing, starting with biological differences even while saying, "Many of these differences are small and there's significant overlap between men and women, so you can't say anything about an individual given this population-level distribution." It's the distance between the words and the actual feeling that stands out—trying to give credence to why women should not be in computing by justifying it with "facts" that are not even proven and even contradictory.

Even women do not expect to see other women in certain male-dominated fields such as technology. I cannot count the number of times I have gotten the raised eyebrow look and questions about working in such a male-dominated field—or how excited I get to run into another woman technologist. Stuck to the floor by inherent biases!

Some think this is because for a woman to advance, a male must either be displaced or their status lessened, a type of zero-sum game where one must lose for another to win. There are a limited number of higher positions with higher wages, and anything that changes the availability of those positions becomes a source of competition. While on the surface people support programs such as work-life balance that try to equalize the workplace for men and women, they will attack or demean those programs if those programs interfere with the availability of the limited resources. Diversity programs make some in the non-target groups feel that the organization is unfairly balanced to promote and advance minorities. As a result, women in these environments may be the subject of the defensive behavior. Benevolent sexism is one covert way that men defensively react in these cases.

It has been ingrained in women that we should not brag about ourselves and our abilities, but how will anyone know what we can do if we don't tell them? Stuck to the floor by our own mindset!

If a woman does find a way to advance, often she hits the glass ceiling. She gets to a certain level and cannot advance past it. We see this in the statistics of how few executives are women in Fortune 500 companies. This contributes to women leaving mid-career out of frustration with the politics of their organization.

If a woman does manage to break through that glass ceiling, she needs to watch out for the glass cliff. It seems that companies will decide to promote a woman when they are in chaos, which sets the woman up to fall right over that cliff. The thinking is that if everything is going well with men in charge, it does not need to be changed. When an organization is struggling, they look to see what they can change to fix it. Bringing a woman in will bring a different perspective and way of organizing and motivating people. But with the company already in crisis, anyone would have a good chance of failure. On top of that, some will then attribute the failure to bringing in a woman, and continue the cycle of looking at women leaders as inferior. It's a frustrating and vicious cycle for women.

Chapter 14

HOW DO WE FIX IT?

We know how gender barriers are impacting women by holding them back from advancement or even joining the STEM field, but the numbers also show that lack of diversity directly hurts those STEM-related companies that are so important to our future. Companies with diverse hiring practices even tend to generate more profits and income, as they are more creative and innovative, benefitting from different viewpoints. Tim Cook, CEO of Apple said in 2017, "Technology will not stay in the lead in the US unless the gender diversity gets materially better. It's just not. It's just not going to happen."[47]

Over 70 percent of technology executives state there is a shortage of skilled and qualified talent. This causes companies to be unable to move forward with innovation, rather than having to keep their staff focused on maintaining existing systems to keep business operating.[48]

47 Holmberg-Wright, & Wright, *Business Education Innovation Journal*, 51-58.

48 Technology Councils of North America. "Persistent Talent Shortage May Hinder Hiring Plans of Technology Companies, New TECHNA Survey Reveals." *PR Newswire*, Web. https://www.prnewswire.com/news-releases/persistent-talent-short-

With the shortages of skilled talent to fill these roles, organizations must rethink their hiring practices. Not having diversity and retainment programs to encourage women and minorities to enter the fields means organizations lose the chance to fill much-needed spots. Companies need diversity in all areas to bring those viewpoints to the table, empowering them to expand beyond current markets.

We have to find solutions to change the dynamic toward women in STEM and put everyone on a more level playing field where skill, knowledge, and talent are the only things that matter. We know that more diversity training is not the right answer. Firms have been steadily increasing emphasis on diversity training since the 1990s, but the numbers are not supporting much change in diversity. It tends to backfire to try to command or shame any bias out of organizations, and it is difficult to determine whether some situations are stereotypical bias or not, as it is not clear-cut. In fact, diversity training tends to really only affect people for a day or two. They know what they are supposed to "say," but it does not change their inherent opinions and tendencies so they go back to their normal quickly. Participants may even feel singled out as the perpetrators causing problems, and this can cause a backlash against diversity.

People tend to react poorly toward negative messaging about their behavior, especially when they are not given direct examples or cannot see how their own behaviors might be considered offensive. A male who thinks he is simply helping out a female coworker does not think he is doing anything wrong—he is just being a gentleman.

---

age-may-hinder-hiring-plans-of-technology-companies-new-tecna-survey-reveals-300005727.html

---

He may not understand that the female does not want or need help, and by not allowing her to handle a situation, she will not be able to learn from it. When you put that male into standard diversity training, he either will fail to see why there is a problem or get angry for feeling singled out. The male will acknowledge that gender stereotypes, judgment, and bias are wrong, but fail to understand his own actions or change his own behavior long-term, as we see with the previous case of the Google employee memo.[49] We do not want men to feel like they must be silent; we want open and honest discussion.

We also have to change the culture of organizations. This will not be easy. In some cases, there may be decades of ingrained behavior and thoughts to overcome. People tend to think culture is always set from the top down, but culture can be changed and influenced at any level because it is not a mandate; it is how people act and interact with each other. If women band together and start small, gathering a few allies from their colleagues, they can start a movement to change perceptions and behaviors at their level. Showing wins and successes will make that movement blossom and start changing the culture at all levels. Let's be and show the change we want to see.

Change Within

The old saying goes that the only thing we can change is ourselves. We cannot force people's ingrained biases and opinions to change, but we

---

49 Edward H. Chang, Katherine L. Milkman, Laura J. Zarrow, Kasandra Brabaw, Dena M. Gromet, Reb Rebele, Cade Massey, Angela L. Duckworth, and Adam Grant. "Does Diversity Training Work the Way It's Supposed To?" *Harvard Business Review*, 9 July 2019. Web. https://hbr.org/2019/07/does-diversity-training-work-the-way-its-supposed-to?registration=success

can change our own attitudes and let that shine out to the world. We can't just talk about how people have biases or believe in stereotypes. That tends to backfire due to groupthink—if everyone thinks that way, then it's okay if I do too.[50] We need to be able to discuss behaviors and attitudes that lead to gender bias so we can start to change people's points of view.

The best way we can start to change the overall situation is to recognize our own negative thoughts and behaviors toward ourselves and other women, and work to weed those out. You have read this book about the stereotypes and biases and judgments that women in the professional world face. How many of those thoughts have you had about yourself or other women? How have you acted or talked about yourself or others that has contributed to the inherent thoughts or opinions of how women should be or act? Have you dimmed your own light because of how others have treated you or how you thought they would view your accomplishments? Have you felt like you are in competition with other women colleagues?

We need to understand and celebrate the strengths that we bring as women and individuals. Everyone brings different strengths and weaknesses to the table. Just because someone does a task or handles a situation in a different way than others, it does not make them wrong. Perhaps they have found a brilliant new way of thinking or acting that could be the way of the future! Women are known to be good at bringing groups together in harmony by having empathy, listening, and

---

50 Adam Grant and Sheryl Sandberg, "When Talking About Bias Backfires", *NY Times*, (December 6, 2014). Web. https://www.nytimes.com/2014/12/07/opinion/sunday/adam-grant-and-sheryl-sandberg-on-discrimination-at-work.html

coaching people to cooperate. This is great news for businesses, so we should capitalize on those talents. We need teams to be able to work together better and be innovative, so this skill should be celebrated.

While this book is geared toward the STEM fields and moving into management rather than physical labor, there are parts of these jobs that may require some sort of physical activity, and we know the stereotype is that women just are not as strong as men. This is where women need to think smarter, not work harder, and bring positive change to the workplace for all. My (younger!) brothers do physically demanding jobs, and both have already stated that they worry about injury or the long-term effects of doing physical labor on things like their joints and spine. They look for ways to keep their bodies from having to take the brunt of the physical labor, just as everyone should; we are not getting younger by the day. Women in STEM have a unique perspective on what is needed to alleviate some of these physical strains and should be called upon to create innovative ideas on ways to remove them.

Remember how women stepped up and took over the manual jobs during WWII? Obviously, women can step up and do manual jobs, and have done so throughout history. So in today's age, with the huge progress in the design of tools, machines, computers, and even robots, there is no reason that a woman cannot handle the same physical labor that a man can. If a job requires physical work, women need to make sure they are physically fit enough to meet the demands of the job. If the job requires strength, she should consider strength training and conditioning before and while taking on the job. If the job requires a person to climb or crawl often, as might be required in stringing cable for a network, the woman needs to assess her agility and flexibility and consider a form of exercise such as yoga to help maintain those skills. Exercise benefits all of us whether male or female, so we should all be doing some form of it. But for women wanting to do jobs that include physical labor, it is even more important they keep their bodies in prime working condition to do the job.

Both men and women need to pay attention to the tools they have available to make their jobs easier and safer. There are ergonomics programs and tools to make sure work is done efficiently and with as little strain on the body as possible. Safety protocols and equipment should be followed to help protect from injury. Safe lifting techniques help save back strain, and regular chiropractic and massage care can help reduce any strain or long-term injury. Stretching throughout the day and after the end of a shift helps loosen those muscles that tighten after hard work, reducing the likelihood of injury while increasing needed flexibility and range of motion. These are techniques everyone should follow, but as women, we should take the lead and double our attention to these concepts to help prevent injury and wear to our bodies. This helps us to show others that we can handle the physical work, and even if weaker, do as good of a job at them as males.

When people try to degrade women's intellect versus men's, we know in terms of general intelligence or IQ that statistically we are equal, and people are more inclined in this generation than previously to agree that women are just as smart as men. However, when it comes to hiring or recruiting for jobs associated with or mentioning the need for high IQ, women are still considered less often than men according to a study done in 2018.[51] We need to push for job postings to remove

51 Bian, L., Leslie, S.-J., & Cimpian, A. (2018). "Evidence of bias against girls and

references to intellectual ability, as this is a known gender bias. Jobs should lay out only the skills needed, and rather than referring to intellectual ability, they should simply state what needs to be able to be solved. Then, the gender of the person will not matter, but only whether they can do it or not. We also need to work to remove any mention of intelligence levels in performance evaluations and advancement opportunities to remove biases that can impede women from advancing their careers or even cause them to leave the field.

Other attributes that women possess that are important to the business world can be highlighted, such as the ability to solve problems working with other team members, which shows openness to learning and teamwork. We need to look at our competency evaluations and make sure these attributes are called out as positives so that women get recognized and promoted into leadership roles. The evaluations used as a basis for promotions and raises should clearly state these competencies to show their importance to the organization, instead of it just being somewhere "on the list" but not actually considered important to the role.

This bias against women being intellectual starts in children, which means they are being conditioned to not think of women as smart or capable of doing intellectual jobs. Women at times play dumb to appease men around them who may feel emasculated by a smart woman. That needs to stop. Men who feel that way are never going to learn that women are not a threat to them if they are not exposed to smart women. And if they cannot change their feelings, they have bigger issues

women in contexts that emphasize intellectual ability." *American Psychologist, 73*(9), 1139–1153. https://doi.org/10.1037/amp0000427

to work through—perhaps with a professional counselor! To dumb down an entire gender to appease male sensibilities is ridiculous. There are plenty of males out there who are not threatened by smart, strong women, and we need to bring our boys up to not fear those women too. We need to praise girls for their accomplishments and praise our boys for supporting women's accomplishments, not just focus on looks for girls and achievements for boys.

Women are also harder on themselves when it comes to intelligence. We seem to think that you are either born with it or not, which is not true. Intelligence can be grown by learning. However, if we teach our girls that if they are smart, they will just understand, and don't teach them how to work for it, they will continue thinking intelligence is innate and unachievable. They will give up before they even try. This carries on to later in life when they are faced with challenges in the workplace. Even if all barriers are removed to advancement, if a woman does not *think* she's capable or smart enough, she will not try for the position. Instead of working toward a stretch goal that requires knowledge or skills that need to be developed, women will set goals they are pretty confident they can achieve with their existing knowledge.

We are starting to see more intelligent women celebrated in movies as well as during Women's History Month, which gives us platforms to share historical figures, but we must highlight our intelligence and not be afraid to show the world in the current day as well. We need to challenge the stereotype that women need to tone down their intelligence to find a man, or stroke their ego to be liked. The more people are exposed to smart, strong women, the less they will fear or feel intimidated by us.

Women need to volunteer for that tough project, even if we do not yet have the knowledge or skills needed, because it shows how much we can learn and adapt. We can show off a skill we have to help others by breaking it down to teach others and showing our knowledge. And this also benefits ourselves, other women and our organizations. Help the next generation by volunteering to work with kids or young adults on learning new skills or just navigating life. You can even mentor college students. I personally did this just to help them learn to navigate the corporate world and how to present their skills to land a job after graduation, and it was very rewarding. It is good for both us and our female colleagues to feel empowered to show intelligence, as well as be role models for our girls.

Specifically, in those male-dominated fields, we may need to find allies among the male colleagues to push women forward who are competent and have the intelligence and skills to do the job. These allies need to listen to what women are facing and the change we want to see, and then help us make changes, without slipping into benevolent sexism and trying to solve the problems for us. Good allies like these can be difficult to find. If we do not have allies in leadership willing to listen to these requests for changes, women may need to advocate for each other in the workplace and not be afraid to speak up and speak out. Celebrate good work and performance, and do not try to downplay it or attribute it to others. Take credit where it is due. Show off those skills that we have or have learned to others on our team or our department. Work on continuous learning, and be curious.

If a woman has skill or expertise in something, we need to ensure her voice is heard. Women need to overcome the stereotype that we do not want to lead by speaking up and leading! It can be hard to break the habit of toning ourselves down or not speaking up in a group when we have knowledge or insight, but it is critical to getting noticed. If a woman wants to be a leader, then she should be able to state that and work on gaining what she needs to move into those roles. Again, volunteer for the project that needs a leader, even if it is outside your comfort zone. Offer to lead meetings or give presentations to groups. We have to promote ourselves; no one else is going to do it. The more women who speak up and move into these leadership roles, the more who are available to help bring up the next generation of leaders.

To be heard in groups or meetings, women need to be ready in advance with what they want to say. Review the agenda before the meeting, and do some research if needed; if there's not an agenda, ask for one. Review not just the topic but the parties involved; look for the politics that may exist behind the scenes. If there is a topic you are particularly passionate or knowledgeable about, prepare what you want to say and be ready to jump in first. Practice what you want to say in advance, whether to yourself or with another colleague or friend. It is even better if you can solicit an advocate or sponsor for your idea in advance to help support your idea in the meeting. Send an advance email to the group and make sure they know you have information about the topic, so you are on the agenda and can gain support and allies. Do not tone down the message with "I think" or "maybe." Be confident in what you have to offer. Keep it short and to the point, as it is harder to interrupt short sentences; plus, it gets your point across better than a long diatribe. Sit as close to the front as possible, or at the table, as women tend to take the back chairs or not sit at the table and thus become secondary to the main meeting. Some say sit in the middle seats at the table, as it is harder to ignore than at the ends of a long table. The power spot is the middle of the long side facing the door. Bring out your confident voice. If you need practice speaking in public, join a group like Toastmasters to learn the skill. Be aware of your body language, practice, and use your confident poses—use your Superwoman stance!

If you are getting interrupted or talked over, you may have to interrupt back. Be kind and thank the person for their thoughts, but state that you want to finish your thought and jump back in. If you have an advocate or ally, they can help direct the conversation back to you. If someone else tries to take credit for your idea, thank them again for bringing attention back to your idea and add any details that were missed. Do not just let anyone take your voice or idea away! You do not have to act offended or shut down; rather, learn to skillfully pull the idea back and take back the conversation. All of this takes practice, so if you can work on presenting to others or join groups to practice your speaking and leadership skills, you will be better prepared when these situations occur.[52]

Women should also watch out for each other, or anyone in a situation who seems to get shut down or doesn't get a chance to have their voice heard. Use those collaboration and facilitation skills to help draw out others into the conversation. If they get interrupted, help direct the conversation back to them until they are finished. You will be helping your own confidence while showing your teamwork skills and helping

52 Beaton, E. (2018, June). "7 practical tips to have your voice heard at the boardroom table." Inc. https://www.inc.com/eleanor-beaton/7-practical-tips-to-have-your-voice-heard-at-boardroom-table.html

another person feel empowered at the same time. Remember, this benefits the organization too. As more ideas are heard, this leads to better outcomes and solutions. These skills require unlearning ingrained habits and culture, so it may take some practice but eventually, as more women speak up, it will get easier overall.

We need women to speak up and be assertive, but to avoid the backlash against being too direct, we need to think of how we frame the message. "Speaking up in forceful, assertive ways is especially risky for women," said Joseph Grenny from the book *Crucial Conversations*. "Emotional inequality is real and it is unfair. And while it is unacceptable and needs to be addressed at a cultural, legal, organizational, and social level, individuals can take control." Women can say things like, "I want to state my opinion very directly" or even just "let me be direct" to call out what their intentions are and not have it seen as negative or aggressive.[53] If a woman does get called too aggressive, she can consider it constructive feedback and look at what words she is using, so she can be less abrasive while still being assertive. But she does not need to lose her basic nature. Women have to be assertive to be leaders and get their voices heard, but we do have to be careful not to genderize the attributes. A woman acting the same way as a man should be labeled with the same positive attribute, i.e., assertive, not aggressive. This is especially important because we tend to judge people more on negatives than positives. Women also need to have our performance evaluated by the same measures as men. We need to get out of the

53 Vital Smarts. "New Study: Women Judged More Harshly When Speaking Up Assertively" (August 5, 2015). Web. https://www.yahoo.com/lifestyle/s/study-women-judged-more-harshly-120000656.html

mindset that we need to hold women to higher standards—and that's woman to woman.

The biggest challenge to women being able to show their talents and speak up to be heard is a lack of self-confidence in their own worth and ideas. Women are conditioned to lack confidence, through the up-bringing that still pushes the agenda that men protect women and are the leaders in thought and voice. This makes it hard for many women to break out of their shells and speak up when we doubt the worth of our own ideas! It is uncomfortable to make these changes and be brave when you feel like you have imposter syndrome—feeling that you should not really be there and someone will eventually find out that you do not know as much or have the skill that everyone else possesses. Finding the confidence to speak up against the fear of being rejected, shot down, or having your idea hijacked is hard.

We are going to have to practice building our self-confidence if we are going to take steps to close the gender gap. We are not doing this just for ourselves, but all those other women and girls who follow us. Women need to understand and celebrate their worth. It is not boastful to be proud of what you do and accomplish and to tell others about it. We need to ask for what we want, just like men. There are many books and studies on how to negotiate for what we want, and most are quick to point out that if you do not ask, you won't get it, and to ask big because you can always negotiate down from that point. Don't be afraid to say no or push back when needed. We tend to want to say yes to everyone and then get overwhelmed. If we have a differing opinion from the group, speak up as it may be the key to a great change.

We also have to get past the unfair and overriding feeling of guilt over all the things we can't do. We can't be perfect moms and wives and professionals all the time. We are all doing the best we can in this world. The concept of having to do it all was built outside of us and thrust on us without our consent or even knowledge. We have to stop feeling guilty and apologizing to ourselves and others when we want to pursue our ideas and make our voices heard. We are allowed to have dreams and ambitions outside of the home and child-raising. We know some things have to wait during different times in our lives but when we are ready, we should have no guilt about pursuing our goals and standing up in the professional world. We also need to get rid of those voices in our head that tell us we are not good enough, or too old, or too young, or whatever the voice is saying to make us doubt or feel guilty. We have to face rejection and judgment and speak up to get ahead in our careers and life.

Remember, people will say dumb things they do not mean or at the moment, but it does not mean that they generally hate women! We tend to revert to our habits, and some of these thoughts about women go back to childhood and the way people were raised. All of us have slipped and said something we did not mean, directed toward a certain person. It is okay to be insulted, and the behavior needs to be addressed, but try to start with a light touch or framing the message calmly to prove your point. We need to use our skills of empathy, nurturing, and teaching to our advantage in these situations. As we change the stereotypes and biases, this behavior can change and we can speak our opinions openly and directly. But for now, we have to take baby steps.

Next, we need to take our experience and skill and use them to help others. We can choose to not look at the next woman in line as competition, but rather as a sister that we hold out a hand to and pull up alongside us. We need to instill confidence in women to speak up and be heard. confirming that their opinions and thoughts are important and valued. We need to recognize that other women may need a confidence boost, or many confidence boosts, to overcome negative opinions and stereotypes within their own minds. It's not enough to tell a woman she has permission to do a task or job. She needs to feel and understand that she has the ability.

Mentoring

Another way to help each other out is mentoring. We know that mentoring programs work whether the mentor is male or female, as they take pride in getting their mentee ahead. Even better is a program with higher-level women mentoring junior women. Men tend to not mentor as many women, perhaps due to that bias of picking people who are more like yourself, or because they fear they will be accused of sexual harassment, or even just do not feel comfortable mentoring a woman. This is why it is so vital to have women leaders mentor other women— we know what we face in the workplace. This is a big part of turning around and pulling the next woman in line up with you.

If the organization does not have formal mentoring programs, offer to start one and recruit others to help. Or pull together more informal meetings, calls, or sharing sessions to discuss issues and solutions that women face. Women leaders need to share their struggles, successes, and insights on how they got to their level. We cannot be afraid to share lessons learned from past failures. We know women need to present as perfect or at least better than average, but we also know people will make mistakes, and mentoring helps the next woman in line learn without having to make those mistakes themselves. The more that people are willing to share, the more others tend to speak up. If you've ever been on those types of sharing calls or meetings, you will hear people agreeing and then adding to the initial conversation if they feel safe and it can help another person.

It does not have to be a formal sit-down mentorship either. In fact, most people tend to prefer an informal arrangement where mentors and mentees establish more of a friendship. I mentored a young woman in her senior year of college for her degree in Information Technology. It only amounted to a few hours per month meeting over coffee or just email, reading her resume, offering tips on how to leverage her past non-technical experience even when applying for technical roles, and talking through upcoming interviews. I also attended her final class presentation to cheer her on. It did not feel like much to me, but she was very grateful and was able to get hired into a good technical role before graduation, and she still keeps in touch! Book clubs studying different topics related to women, business, and leadership with guided discussion can also be helpful to get people engaged. General discussion forums where members can bring up issues and ask for guidance or thoughts from each other are useful too.

Many times, the mentoring is about networking and navigating politics – who and what you know that is coming up that can benefit the mentee. It can be hard to find out about upcoming openings or skills gaps unless you are in certain circles at an organization, so having someone help you navigate those situations is immensely helpful. Understanding the business or different areas that a woman can explore will help round out her skills and education. Just exposing women to other ways of thinking or doing tasks can help improve their own performance and potential advancement!

These programs also benefit the mentors as they gain insight into what employees are facing in the workplace and where skills might be short. It helps with succession planning to have people being developed and learning what their mentors know. The mentor may even find a gold nugget of talent who just needs nurturing and confidence to blossom and take the organization forward. It will also grow the network of women who are connected when those future leaders mentor other women, and stop the inter-gender conflict to try to stand out against each other.

When mentors find those women with an exceptional talent for an area, they need to become sponsors—going above offering advice to actively advocating for their candidate. A sponsor is willing to go to bat for you even when you are not in the room; they will drop your name for ideas and opportunities that you would otherwise not know about. Males find sponsors to help tell others about their skills and get them into consideration for top roles; women need to find and sponsor other women for those roles. If the hiring committee only sees male candidates, they are going to choose a male. If there are good female candidates, it levels the pool somewhat, and more females will be considered for top roles.

Changing Perceptions

When people do not see women in various male-dominated fields or roles, they continue to believe that women are not interested or are not good enough to be in that position. It's similar to a girl wanting to do a sport like football, but not very many other girls do it and so they don't pursue it. I am happy to say that view is slowly changing; my own daughter is playing high school football, and she is a defensive tackle! However, we had to specifically ask the coach for information about playing because the school only sent it to the boys—though after our request, it was sent out to all students, another unintentional slight to girls. In the same way as having more girls participate in sports, we also need numbers and visibility of women in STEM fields to change the minds of our own gender. That will also chip away at those stereotypes and biases by exposure. By getting more women into leadership roles, it will show that women have the desire and are good leaders.

We need to be brave and throw our hat in the ring for challenges and promotions, even if we feel that we are not as qualified as someone else. Women tend to downplay our talents, and if we are not 100 percent a fit for the job, we think we should not try, whereas males will try even if they only have most of the qualifications. We need to learn to self-promote and take chances and risks to get into those higher roles.[54] Mentoring and sponsorship from other women who have faced these challenges will help with the self-confidence to put ourselves forward. As Cheryl has learned from working as a Nurse Practitioner (NP) in an acute care hospital setting for over thirty years:

The NP can work independently, but most of our life as a nurse we followed physician orders. Now, many do not know how to stand up for their own worth in a man's world... Respect is earned, and I am in a

54 Christina Pazzanese. "Women less inclined to self-promote than men, even for a job." *The Harvard Gazette,* 7 Feb. 2020. https://news.harvard.edu/gazette/story/2020/02/men-better-than-women-at-self-promotion-on-job-leading-to-inequities/

system that allows us to drive change. I have taken the lead role in this change as the APC system lead. Standing back to feel the victim of stereotypes does not improve the outcome or change one's mind. You have to learn to stand up and be noticed.

Networking Circles

As more women enter into male-dominated fields or leadership roles, we need to support them and not let them feel alone. The more we work together, the more opportunities will open up for all women. We know that being competitive with each other and ostracizing women who are advancing does not work; it just makes the prospect of advancing to those roles less enticing for good female leaders. It also lends to the stereotypes that women do not work well with other women. Social networks are important in the workplace; we know that men have their networks with contacts and how important it is in their advancement. As women, we know we face more stereotypes and biases, so having a good network of women to share experiences with helps women leaders not feel so alone and excluded.[55] Women in these networks can act as mentors and help sponsor each other along with sharing information and experiences.

To make these networks successful, we need to put time into them. We need to be available for each other. If a woman has a question or needs assistance, we need to be available to support each other. We need to share information about potential opportunities and help each

---

55 Brian Uzzi. "Research: Men and Women Need Different Kinds of Networks to Succeed." *Harvard Business Review*, 25 Feb. 2019. https://hbr.org/2019/02/research-men-and-women-need-different-kinds-of-networks-to-succeed

other succeed. I have worked at several organizations that had Women's Team Member networks in addition to other diversity and inclusion networks. Through these networks, women can have safe discussions and gather thoughts and information from other women in the organization. It helps to have diverse opinions from different levels and departments to share information that otherwise might not be known or accessible.

If your organization does not have an official women's group, it is not hard to start one. Can you gather some women to do a lunchtime meeting once a month? Can you set up coffee chats? If you do not have time during the workday, can you set up an after-work meetup or even a before-work meetup? Even if the leader does not have a lot of time to dedicate to mentoring, they may have an hour they can give if you find them a topic that they are passionate about or a topic they want to share. Can you get people to discuss what they do in their particular department? Find a good book about advancement, leadership, or women's issues and discuss it. Offer question and answer sessions where women can ask whatever they want in a safe space and share ideas. Small starts will help get people engaged and excited to join and help.

Celebrating Successes

Women's groups are a great way to discuss and celebrate successes. Women should feel safe and able to share good things that they have done and celebrate wins without feeling like they are bragging. Sharing successes in a safe group is good practice in learning how to share with others. If someone is struggling to think of what strengths and accomplishments they have, other women can help them think through their achievements and skills. They can also practice telling each other about what they think they do well. We also need to learn how to celebrate other women's success and not devalue their accomplishments or make it a competition. Having discussions on how each woman's wins benefit all women will help drive out negative thoughts.

Outside of these women's only groups, we need to be calling out and celebrating women's successes in our workplaces so that everyone hears about them. If a colleague has a successful project or gets a promotion, rather than trying to take away from that success to keep ourselves looking better, we need to be better at publicly discussing and celebrating our fellow coworkers. Many women will attribute their success to their team, which is fair, but her contributions need to be specifically called out as well. If you see another woman who does something particularly well, tell her how much you appreciated what she did. If a woman has a skill that you need help with, ask her to assist you or teach you. If a woman advances, even if it was a promotion you were hoping for, congratulate her and tell people about her qualities and talents that made her right for the position. If gossip starts about her advancement, stop it and point out what she has done and why she deserves and earned the role. We need to stop our own negative thoughts about other women and what they accomplish, and stop others from continuing negative messages.

This is also why we need sponsors, specifically female sponsors, available to speak about women's skills when opportunities arise. We need that person who is able to speak on our behalf when we are not present in those meetings and discussions that men get promoted in.

We need sponsors who can describe what women have done well and how they could stretch into new roles or leadership. It should not seem an anomaly that a woman is put up for special projects or roles, or brought in just to balance the gender equation. Women should be discussed based on their skills and potential just like the men, and we need sponsors willing to normalize that discussion.

Chapter 15

OUR RALLYING CRY

This is a rallying call for women in the STEM world. We need to make changes to make it better for ourselves and the future generations of women we need in the STEM fields. The only person you can change is yourself. We can't change people who still see us as inferior and want to put us into a mold of the "correct" way a woman should act. We can't change history. What we can change is how we see ourselves and each other, how we express ourselves and our worth, and how we treat ourselves and others. As a tribe of smart, strong, women, we are half of the people in the world and we have strength in numbers when we stand up and say, "Enough." And after we pull ourselves up, we need to turn around, help out our sisters, and pull each other up.

To be clear, this is not saying that every woman wants to be a CEO. This is about understanding and changing our thoughts, behaviors and judgments about ourselves and others in whatever they choose for their lives. It's about removing barriers that are simply based on our gender. It is about making the shift from what we have been told we should do or think into an open-mindedness so we can make our own decisions about what is best for us. It is about support and understanding and even empathy for each other. That is how women will take their true spot in the world.

The STEM fields are tough. They require discipline and innovation. They are also what will take us into the future. If we can get organizations and people to think differently about women and our contributions, we can tap into a wide market of intelligence and new ideas. We need to figure out how to make these fields more open and inclusive for all. The effort will be worthwhile when we see a world where girls and women can join exciting fields and feel like they are equal partners and contributors. Where it is not abnormal to see a woman leading a group including other strong, smart women and making monumental changes to the world. Where our girls are not afraid to be smart and strong and try anything they want to put their minds to. Where we do not have preconceived notions and ideas of how everyone should look, act, and think, and we strive to learn about the person and their skills before we judge their capability.

This will take us into the future. This is how we pull each other up.

For more information or
to purchase copies of this book,

visit the author page for Heather Graham

at www.bushpublishing.co.

www.ingramcontent.com/pod-product-compliance
Lightning Source LLC
Chambersburg PA
CBHW071435210326

41597CB00020B/3800